THE WILL TO CREATE

THE WILL TO CREATE

GOETHE'S PHILOSOPHY OF NATURE

Astrida Orle Tantillo

UNIVERSITY OF PITTSBURGH PRESS

ST. PHILIP'S COLLEGE LIBRARY

B
2987
.Z7
T36
2002

Published by the University of Pittsburgh Press, Pittsburgh, Pa., 15261

Copyright © 2002, University of Pittsburgh Press

All rights reserved

Manufactured in the United States of America

Printed on acid-free paper

10 9 8 7 6 5 4 3 2 1

LIBRARY OF CONGRESS CATALOGING-IN-PUBLICATION DATA

Tantillo, Astrida Orle.
   The will to create : Goethe's philosophy of nature / Astrida Orle Tantillo.
      p. cm.
   Includes bibliographical references and index.
   ISBN 0-8229-4177-5 (alk. paper)
   I. Goethe, Johann Wolfgang von, 1749–1832—Contributions in philosophy
of nature.   2. Philosophy of nature—History.   I. Title.
   B2987.Z7   T36 2002
   113'.092—dc21

                                                          2001006539

ST. PHILIP'S COLLEGE LIBRARY

For Steve

CONTENTS

PREFACE

THIS BOOK REVISITS Goethe's status as a natural philosopher. Specifically, it investigates the principles behind his conception of a will-driven nature and analyzes the significance of these principles for such philosophical issues as objectivity, scientific method, theories of perception, aesthetic judgment, gender categories, and the status of natural law generally. Although studies have been done focusing on the historical and cultural influences upon Goethe's science,[1] the relationship between his scientific and literary works,[2] Goethe as an historian of science,[3] and individual scientific works,[4] this study is the first comprehensive examination of Goethe's natural philosophy across his scientific corpus.

In the twentieth century, thinkers (like Ludwik Fleck and Thomas Kuhn) revolutionized the way we think about scientific truth, methodology, and discoveries. Fleck's "thought-styles" and Kuhn's paradigms emphasize that scientific "facts" do not arise or exist in a vacuum, but are the products of historical contexts and points of view, individual and collective psychologies, and received beliefs that have been passed down from earlier generations. That Goethe raised these issues one hundred and fifty years earlier is perhaps a surprising notion to those accustomed to viewing him primarily as a poet or a canonical and conservative icon. Yet, as I argue during the course of the book, Goethe indeed questioned some of the most basic tenets of his own time, tenets that until quite recently also went unchallenged in our own. He approached these questions, moreover, from a completely different perspective and had completely different goals in mind.

One of the main reasons that Goethe is not better known as a natural philosopher is that he is perceived to have failed as a scientist. First and foremost, he is generally viewed to have been on the "losing" side of a scientific controversy: he tried to challenge Newtonian physics and methodology and substitute an alternative, more qualitative approach. In addition, many of his scientific observations were viewed as the products of poetic fancy. Until the advent of chaos theory, his creative conception of nature—a view that stresses that the smallest of organic changes or the most minor of diversions in patterns may create new organisms and phenomena—was seen as nothing more than an artist's imaginative (and ungrounded) portrayal of nature. And because his scientific works at times employ poetic terms, when these texts are analyzed at all, they are more often than not examined as literary, rather than scientific or philosophical, works.

A final difficulty in examining Goethe as a philosopher is that many have treated his science as a kind of mysticism or religion. While many of Goethe's earlier readers attempted to find religious overtones within the natural writings of their hero, the propensity of treating Goethe's scientific texts as mystical rather than philosophical is best illustrated in the works by Rudolf Steiner (1861–1925) and his followers, the anthroposophists. The anthroposophists, who often write on Goethe's science, tend to look within his texts for messages of personal/spiritual guidance and fulfillment.[5]

This book does not set out, as others have done, to rehabilitate Goethe as a scientist, make claims about the particular influences that his scientific works may have had,[6] or argue for the rightness or wrongness of particular results.[7] Although Goethe may have made contributions in different scientific fields, his works have not exerted a profound impact upon the scientific community. However, what is significant about Goethe's scientific texts is the philosophic principles that underlie them. These principles illustrate how Goethe attempted to break with past traditions and formulate an alternative philosophical approach. He rejects some of the most ingrained scientific doctrines of his time, including teleology, the chain of being, preformation, and epigenesis. In their stead, he presents a more dynamic understanding of nature, where nature's parts, whether animate or inanimate, act according to their own impulses. These impulses, moreover, are different than what his contemporaries viewed as instinct and the formative drive. Individual entities in Goethe's conception of nature possess such a strong will that at times their choices and desires bring about changes in their own forms and modes of existence. Nor does nature act as it does due to divine intervention or divinely inspired ordered rules. The behavior of nature is largely explained by its will (perhaps foreshadowing the later Nietzschean sense) to create.

✤

This book could not have been possible without institutional support. My year spent as a fellow at the Institute for the Humanities at the University of Illinois at Chicago enabled me to finish large portions of the manuscript. Archival work in Germany was made possible by the Institute's travel funds as well as a fellowship from the Weimar Classics Foundation. In Weimar, thanks to this foundation, Dr. Lothar Ehrlich, and Dr. Gisela Maul, I was able to work with Goethe's own scientific instruments, examine his copies of seminal scientific texts, and use the extensive collections of the Herzogin Anna Amalia Library. The staff at the Anna Amalia library and the photography archives were extremely helpful.

Portions of this book have appeared in earlier versions and in slightly different contexts. Permission has been kindly given from the *Goethe Yearbook* to use sections of "Deficit Spending and Fiscal Restraint: Balancing the Budget in *Die Wahlverwandtschaften*" (7 [1994]: 40–61), from *Seminar* to use sections of "Polarity and Productivity in Goethe's *Die Wahlverwandtschaften*" (36 [2000]: 310–25), and from *Eighteenth-Century Life* to use "Goethe's Botany and His Philosophy of Gender" (22 [1998]: 123–38).

I thank the individuals who gave me very helpful comments or advice, including Hans Adler, Sander Gilman, Helga Kraft, Amy McCready, Imke Meyer, Mary Beth Rose, Susanne Rott, Marian Sperberg-McQueen, Wilhelm von Werthern, and my fellow colleagues at the Humanities Institute. Sem Sutter, of Regenstein Library at the University of Chicago, has provided invaluable help throughout my many years of research there. I especially, however, thank Gabrielle Bersier, John F. Cornell, Karl J. Fink, Daryl Koehn, and Dennis Sepper for their extensive comments and support. I am particularly grateful to John, who, as my first teacher, opened a new world for me. My interest in Goethe began in graduate school, and I greatly appreciate the help and guidance that I received from Saul Bellow, Werner Dannhauser, Peter Jansen, Leon Kass, and Karl Joachim Weintraub. Finally, I thank my husband, Steve Tantillo, who helped me on every aspect of this project. I could not have written the book without his insights, our long conversations, and his extensive comments on the manuscript itself. His contributions cannot be measured, and I dedicate this book to him.

FA   *Sämtliche Werke.* Ed. Hendrik Birus et al. 40 vols. Frankfurt: Deutscher Klassiker Verlag, 1985 –.

GA   *Gedenkausgabe der Werke, Briefe und Gespräche.* Ed. Ernst Beutler. Artemis Ausgabe. 24 vols. Zurich: Artemis, 1948–1954.

HA   *Goethes Werke.* Ed. Erich Trunz. 9th edition, 14 vols. Hamburg: Christian Wegner Verlag, 1950–68.

LA   *Die Schriften zur Naturwissenschaft.* Ed. Rupprecht Mattheai, Wilhelm Troll, and K. Lothar Wolf. 2 parts. 11 vols. Weimar: Hermann Böhlaus Nachfolger, 1947 –.

Unless otherwise noted, all translations from Goethe's works are taken from *Goethe's Collected Works,* 12 vols. (New York: Suhrkamp, 1983–89). Citations are listed according to volume number and page of this edition. Emendations to the translations are marked by an asterisk (*).

THE WILL TO CREATE

INTRODUCTION

IT MAY SURPRISE those who know Johann Wolfgang von Goethe (1749–1832) as a literary figure to learn that he believed that his greatest accomplishments were in the fields of science and natural philosophy. He quite seriously thought that his literary works, including *Faust,* were insignificant when compared to his scientific studies (FA, 2, 12: 232, 320). During his lifetime, he variously worked in the fields of physics, botany, morphology, zoology, geology, and meteorology. Time and again, he attempted to show the seriousness of his scientific endeavors by writing essays that outlined his qualifications or justified particular results (FA, 1, 24: 399–438, 732–52). And although his scientific works were largely ignored, he predicted that they would bring about a revolution, i.e., that they contained a critique of Newton (1642–1727) and Enlightenment science that would start a new era in science (2 May 1824, FA, 2, 12: 116).

Goethe contended that the theories of men like Newton and Descartes (1596–1650) were impeding an understanding of nature. In his mind, Enlightenment science had become so powerful that people were no longer free to pursue scientific ideas outside of an accepted canon. He therefore wanted to "lift the veil" from Newtonianism, which had, by virtue of Newton's personal power and authority, long "obstructed an unprejudiced view." True nature, he argued, had to be brought to light by sweeping away "old fallacies" (FA, 1, 23, pt. 1: 14). It became Goethe's mission, which he pursued with a rhetorical zeal that often clouded his own message, to show the shortcomings and failures of modern science and to propose an alternative approach of studying the world. He wanted to replace Newton's and Descartes's mathematical, deductive, and analytical constructions with phenomenological ones. Science was to be not only for an elite class of mathematicians, but for anyone who was willing to observe nature closely.

There is an obvious gap between how Goethe viewed himself and the success of his scientific project. While there have always been a handful of philosophers and scientists throughout the last two hundred years (including Hegel, Schelling, Darwin, and most recently Feigenbaum, Libchaber, and Sacks) who have supported and championed aspects of his science, his scientific works and natural philosophy have not found widespread support, and some of the scientific claims that he held most dear have been proven to be simply wrong. Indeed, several of Goethe's readers have viewed his hyperbolic statements and even his interest in science as symptoms of illness. To most of his readers, both past and present, his scientific works are poetic endeavors that at best enhance one's appreciation for certain details of nature. And no matter how he himself valued his scientific works, it also seems a distortion to disavow *Faust* in favor of *The Theory of Colors* (*Zur Farbenlehre*, 1810). Hermann von Helmholtz (1821–94), one of the most famous critics of Goethe's *Theory of Colors*, places Goethe's claims about the importance of this work alongside Richard Wagner's declarations that his verses were superior to his music ("Goethe's Presentiments," 396). According to Helmholtz, Goethe was incapable of thinking like a scientist ("On Goethe's," 7–14; "Goethe's Presentiments," 395–98). And, instead of seeing Goethe in the role of a revolutionary, most scholars, especially since the 1960s, have seen him as a figure epitomizing the canon. For them, Goethe stands for conservative values, whether in the context of classical science, aesthetics or, as has most recently been argued, oppressive political regimes.

A close look at Goethe's scientific texts, however, provides a different perspective on his thought. Many of his ideas, especially those involving concepts of truth and objectivity, are quite radical. Goethe consciously saw himself as part of a long line of revolutionary natural philosophers and tried to emulate their models and rhetoric. His first publication on color, *Contributions to Optics* (*Beiträge zur Optik*, 1791–92), intentionally followed a Baconian type of inductivism (Sepper, *Contra*, 17, 41–43). He also took his cue from Descartes's *Discourse on Method* (*Discours de la méthode*, 1637) in defining the scope of his *Theory of Colors*. Goethe turned to Descartes's rhetoric when he wrote about discovering the "right path" and the necessity of "razing, dismantling, and demolishing" the structures left by his predecessors. But where Descartes desired to clear away the impediments of religion and Scholasticism, Goethe viewed the greatest obstacle as the mathematically based methodology that Descartes himself had promoted. For Descartes, the problem with philosophy, as with the sciences, was that people could never completely agree on the truth. He sought the certainty that could be obtained with a mathematical and mechanistic view of the world and predicted that the future of science lay in the disembodied subject and mathematics. Nature was simply matter operating according to regular laws and principles. Cartesian certainty, moreover, could be discovered apart from nature, through the exercise of one's mind.

In contrast to Descartes, Goethe held that the most essential aspects of nature (whether animate or inanimate) are those that are not easily quantifiable. As a consequence, his scientific studies were more interested in irregular forms, changing patterns of behavior, and the influence of particular contexts upon individual phenomena. He discounted the possibility of ever discovering permanent truths about nature because nature is constantly changing and redefining itself. Arguing that nature could only be understood in action and in dynamic interaction with all of its parts, he assailed modern science's propensity to conduct experiments primarily in controlled settings. In addition, he maintained that the best scientists were those who studied nature from a variety of approaches and perspectives, while the worst were those who saw the world through only one perspective, whether, in his words, through the eyes of a "geometer" or "mechanist" (letter to Schiller, 13 January 1798, FA, 2, 4: 479–80).

At the core of Goethe's alternative science is a concept of the observer that is strikingly modern. Whereas Descartes advocated a complete separation between the thinker and the world, Goethe focused upon the individual's relationship with the world. He argued that it is impossible to conduct truly objective experiments because all scientists, whether they admit it or not, theorize each time they examine the world (FA, 1, 23, pt. 1: 14). They therefore could never completely abstract themselves from their own agendas and theories. Good scientists had to approach nature with self-knowledge if they were going to differentiate between their own subjective contributions to the experiment and nature's own activities. Instead of making absolute or objective claims, they would examine their own theoretical perspectives and prejudices while conducting research. This approach, according to Goethe, circumvents some of the difficulties of subjectivity by openly acknowledging them.

ST. PHILIP'S COLLEGE LIBRARY

Similarly, when assessing the scientific findings of others, it was important for Goethe to know something about the personal histories of the researchers, as their biographies could enable one to see some of their particular influences and prejudices.

By focusing upon the subjective prejudices of scientists, Goethe acknowledged that all scientific methods are, to a degree, cultural constructs. Nevertheless, he still maintained that some scientific approaches are better than others, i.e., that certain types of standards, albeit dynamic and changing ones, exist. He makes this claim because he believes that nature is always changing and that scientific approaches to it constantly need to adapt. In other words, new scientific approaches may need to be developed as natural forms themselves change. In his scientific works, he attempted to devise a methodology that would reflect a vital and changing object of study. Instead of formulas, he presented dynamic principles that describe the fluid and sometimes even unpredictable activity of nature. His scientific works paint a chaotic, yet patterned, portrait of nature.[1]

Goethe's objections to the methodology of modern science were not isolated to the use of symbols and formulas. He also questioned the adequacy of language within scientific research.[2] Language (or the written text) had to be accompanied by visual texts or experiences in order fully to display the complexities of nature. For example, in the preface to his *Theory of Colors,* he compares his scientific experiments with a dramatic play: only half of the experience resides in the written text (FA, 1, 23, pt. 1: 18). We can only understand the full impact of the experiments if we have them in front of our eyes and experience them, emotionally and visually, as we would a dramatic staging. Consequently, his own drawings accompany many of his scientific essays, and he worked with artists and technicians to prepare visual materials to accompany several works, from the colorful charts for his *Theory of Colors* to the detailed watercolors for his "Metamorphosis of Plants."

Finally, Goethe emphasized the importance of studying nature because he believed that human beings erroneously place themselves above or outside of nature. Like many environmental ethicists today and the American Transcendentalists, such as Ralph Waldo Emerson (1803–82) and Margaret Fuller (1810–50) who acknowledged his influence, Goethe argued for an integrated view of nature.[3] Scientists, by denying

ST PHILIP'S COLLEGE LIBRARY

their connection with the natural world, had missed certain aspects of nature. Whereas Goethe's scientific predecessors and contemporaries argued that the absence of the intermaxillary bone in the human skull separated human beings from the apes (FA, 1, 24: 16), Goethe was rightly convinced that no such separation existed. He began his search for this bone in the human skull because he wanted to prove the marvelous (bewundernswürdig) continuity of nature (FA, 1, 24: 24). He also argued that human beings could learn about their own natures by observing nature at large. In "Maxims and Reflections," he contends that all existing things are related in analogous ways (FA, 1, 13: 46, no. 1.293) and that nature, whether writ large or small, is the same (FA, 1, 13: 401, no. 6.34.1). He therefore advocates studying colors as the actions and passions (Taten und Leiden) of light in his *Theory of Colors* and presents a chemical theory to explain human behavior in his novel, *Elective Affinities* (*Die Wahlverwandtschaften*, 1809). To Eckermann he reported that he learned about human nature through his scientific studies and that human politics follow the same patterns as organic development (13 February 1829), whereas to Charlotte von Stein he wrote that the laws explaining the development of plants may be applied to all living things (Rome, 8 June 1787, FA, 2, 3: 305–6).

At the heart of Goethe's scientific portrait of nature are several principles. These describe the inner drives of animate and inanimate nature and stress the nexus of relationships among all natural things. He refers to these principles whether discussing plant metamorphosis, colors, cloud formations, or comparative anatomy. Because any single part of nature may at any time exert its will and change the structure or the pattern of the whole, Goethe's principles do not use mathematical formulas, but instead give visual examples or case studies that highlight the dynamic relationship of nature's parts. Nor do his principles encourage the analysis or the dissection of natural parts or activities. Rather, his principles, which are themselves closely related, focus upon the relationships among parts and wholes, animate and inanimate objects, or nature and human beings. They therefore represent, on the one hand, Goethe's attempts to uncover a philosophy equally applicable to human beings as to nature as a whole. On the other hand, however, they also highlight some of Goethe's weak spots as a scientist: his rejection of analysis, dissection, mathematical formulas, and almost all scientific instruments enhancing perception, including microscopes, telescopes, and even glasses.

An examination of the interrelationship of his main scientific principles, especially in respect to his views on gender and growth, brings to light a more brutal image of nature than is generally associated with Goethe. Many who write on his science stress the orderliness and patterns of his scientific views, whether his favoring of neptunism over vulcanism (e.g., Muschg, 58; Wells, 72), his portrayal of nature's development according to gradated steps (e.g., Boyle, 1: 594; D. Kuhn, 6–12), or his use of archetypes (e.g., Helmholtz, "On Goethe's," 2–3; Wells, 18–21). A close look at his scientific texts, however, reveals that he is also interested in nature's disorderly manifestations: the retrogressive, the monstrous, the combative, the irregular, and the revolutionary. This is not to deny that Goethe often writes of nature's regular processes (such as the metamorphosis of a regular plant), but enough emphasis has not been given to nature's aberrations within his works. Goethe turns over and over again to those natural entities that do not follow the norm. These irregular forms, often brought about when one part of an organism destroys the regular hierarchy of the whole, allow Goethe to discover what would otherwise remain hidden (e.g., FA, 1, 24: 111, no. 7; 611, 779). Thus, his scientific texts describe a view of nature that is much less patterned and regulated, and much more fluid, than is often portrayed.

One chapter is devoted to each of Goethe's four main scientific principles: polarity, Steigerung (intensification), compensation, and competition.[4] A theme that unites all four—no matter what aspect of nature they address—is nature's creative drive and powers. Each principle describes nature's will to create, evolve, struggle, transform, and metamorphose. If nature creates laws for itself, it also oversteps those laws and creates new ones. Goethe therefore uses his scientific principles, not as a means to reduce the realm of nature to set laws, but to describe, from several different perspectives, the creativity that he sees present in nature.

Chapter 1 explores Goethe's principle of polarity. Polarity, for Goethe, is one of the main "driving forces of nature" and serves as the "language of nature." He argues that the entire universe is comprised of opposing entities that are breaking apart and rejoining and in a state of constant flux. Like many of the most important twentieth-century philosophers, Goethe was influenced in this view by the pre-Socratics. However, where many today see polarities primarily in terms of contradictions that serve to collapse the meaning of language and art upon themselves, Goethe saw polarities as the source of creativity and meaning. Unlike earlier philosophers, he does not rank one polar side above the other, because both are equally necessary to the creative process. Accordingly, "ideal" forms arise with a balanced union between two opposites, whether those opposites are represented by body and soul or action and passion.

Goethe's creative view of polar interactions has a significant influence upon his scientific methodology. Within his scientific texts, his method often incorporates two opposing theories within it. For example, he variously employs aspects of epigenesis and preformation, or atomism and dynamism. (He also claims, in theory, to want to include aspects of both analysis and synthesis.) Most importantly, however, his understanding of polar relationships causes him to question the reliability of the senses and to reject the possibility of an objective observer and objective knowledge. Scientists, whether they are aware of it or not, are always involved in a polar relationship with the objects of their study. As a consequence, the objects influence the subject (sometimes even in pathological ways), the subject always brings a particular perspective to the study of the object that ultimately will influence the results, and the subject, involved in its own polar relationship with the object, may "create" phenomena in order to have a "whole."

Chapter 2 examines Goethe's most hierarchical and hence most controversial principle, that of Steigerung. Through this principle, each of nature's individual parts struggles to ascend to greater complexity and perfection. This concept has been made most famous as the impelling force behind Faust's actions and his ultimate ascension into heaven. It is, according to Goethe, also the operative force behind a plant's metamorphosis as it grows and its organs become more specialized and is responsible for the formation of certain intensified and highly symbolic colors. Goethe uses this principle to refute the botanical theories of both Carl von Linnaeus (1707–78) and Caspar Friedrich Wolff (1734–94) as well as using to it to support Johann Friedrich Blumenbach's (1752–1840) concept of the formative drive (Bildungstrieb). Where some have viewed Goethean polarities as evidence of his democratic tendencies (Burgard, 215–26), others have turned to Steigerung as an example of

Goethe's elitism. In contrast to Aristotle, however, Goethe did not believe that ends are static or predetermined. Steigerung could never represent the quest for absolute knowledge because nature undergoes so much flux that knowledge of it can never be stable. Instead, each new generation of natural entities must be studied anew because their ultimate or final goals or ends are always changing. Like Faust, natural entities are never self-satisfied and constantly struggle to achieve new heights. Whereas Aristotle limited nonhuman striving to procreation (*De Anima,* 415a27–b3), Goethe viewed nature as much more creative. He points to beautiful but useless features and forms as evidence of nature's drive to flourish beyond simple existence. Therefore, Steigerung, on the one hand, is hierarchical because it describes how nature overreaches itself and how entities strive to become more complex and articulated. On the other hand, it is also democratic. Goethe does not limit striving to higher organisms or even to animate ones. Rather, every single part of nature, whether animate or inanimate, large or small, may participate in striving or self-overcoming.

This chapter further investigates Goethe's views on teleology. He repeatedly and forcefully rejects teleology as a means of studying nature. However, his notion of Steigerung, which emphasizes increasing complexity and perfection, may seem to imply a tacit acceptance of some version of it. By carefully distinguishing Steigerung from a classical understanding of teleology, however, one may see how he attempts to oppose teleology while embracing the creative hierarchy represented in Steigerung. Such an exposition, in turn, may help explain why he has been variously classified as a conservative and a liberal, an adherent to classicism and romanticism, and an Aristotelian and a (pre-) Derridean.

Chapter 3 continues the discussion of creative formation and organic development by turning to Goethe's principle of compensation. It links Goethe's discussions of animal types within his essays on comparative anatomy to an evolutionary theory. His principle of compensation further outlines how nature's most creative and "beautiful" acts arise when it is faced with limits. Within his anatomical and zoological essays, Goethe uses the economic language of a balance sheet in order to discuss how animals differ from one another both within one species and among disparate ones. For example, he describes how a snake "pays" for its long body at the expense of its feet and how a frog's long legs account for its stout frame. Goethe further discusses individual examples of "good" and "bad" expenditures in those cases where certain animals evolve and change their forms and way of life. This judgment is based upon the organism's capability to move freely, a concept that he then closely relates to beauty generally. The most free animals are also the most beautiful and vice versa. Several of his essays and scientific poems reveal that Goethe also uses this kind of economic terminology to discuss the potential for creativity of inanimate entities and human beings. This chapter includes an investigation of the scientific methodology that Goethe develops in conjunction with the principle of compensation. He postulates the necessity of turning to artificially created nomenclatures and types in order to discuss and study a nature that is always in flux.

Unlike polarity or intensification, this Goethean principle is seldom addressed in secondary literature (Graham, 153ff.; Jackson; Lenoir, "Eternal Laws," 24; Tantillo, "Deficit," 40–41; Wells, 19–23), although it is central to both his scientific and literary works. Whereas several scholars have argued that Goethe did not believe in evolution,[5] a close examination of this Goethean principle presents strong evidence that he promoted a view of evolution that was quite different from Darwin's. Nature is not spontaneous, but will driven. Goethe's principle emphasizes the will and the drive of individual entities to change their own forms and functions, as when a sloth evolves into an apelike creature to gain more freedom of movement or a type of rodent strives toward creativity in its architectural endeavors. This principle also describes how inanimate things (such as clouds) have their own creative wills and characters.

Chapter 4 examines Goethe's philosophy of competition—especially as it relates to issues of gender and reproduction. His concept of competition illustrates the prevalence of the creative urge throughout nature. Within his morphological texts, he expands the notion of reproduction to include growth. He considers each organism not as a single entity, but as a community of parts, where each part strives to reproduce itself. His theory of reproduction presents a political understanding of the relationship of organic parts to the whole that focuses upon the competitive (and creative) aspects of the reproductive processes. This competitive battle becomes especially acute in his treatment of the masculine and feminine tendencies within plants. According to Goethe, the two opposing sides of the masculine and the feminine can successfully join only if they have both undergone a process of intensification. Their natural relation to each other, however, is a state of war and competition where one force tries to dominate the other.

While historians of science have examined several eighteenth-century scientists and argued that theories of procreation often mirror people's social attitudes toward women (Farley, 3) or provide a context for aspects of intellectual history (Roe, 1–20, 148–56), Goethe's views of gender and procreation in his botanical works remain largely unexplored.[6] Yet his scientific works shed new light on the gender debates of his time. First, the fluidity of his gender categories is unique, and, second, he at times subverts the traditional assumption of the supremacy of the masculine.

In his botanical works, Goethe argues that if we are to understand gender at all (in human beings as well as in plants), we must begin by studying the androgynous qualities of plants. He presents a theory of gender based not upon reproductive organs, but upon actions and characteristics. He ascribes to the plant masculine and feminine tendencies that exist independently from sexual organs. He gives precedence to neither the masculine nor to the feminine, but describes their relationship in terms of a struggle for power. Indeed, nature is so free to express its will that in some cases an organism may change its gender throughout its lifetime. Thus, within one individual plant, at times the masculine may predominate, while at others the feminine gains control. Within these botanical works, Goethe often draws analogies between plants and human beings and their endeavors. He likens the competing forces of the plant to the relationship between two human beings and between the competing fields of science and poetry.

❧

A study of Goethe's scientific texts demonstrates that the same thing that makes his science so interesting is also that which makes Goethe his own worst enemy. Because he sets out to critique mathematics, analysis, and reductive approaches, he appears to be antiscientific. Moreover, his scientific method, which is largely based upon observation, has led critics, including Walter Benjamin, to question whether there was anything more to Goethe's science than lists of visual experiments (314–16). Further, because he rejects strict systems and poses in their place dynamic, changing principles, his natural philosophy appears random—especially within the context of the more systematic philosophers of his day. Even Goethe's efforts to undermine the authority of the Enlightenment led to a basic misunderstanding of his own science. His attempts to popularize science included the writing of didactic poetry to explain more complicated scientific principles to a wider audience. As a result, his scientific project was received more in the light of a poet's musings than as a serious thinker.

In this book, I address all of the aforementioned issues within the context of Goethe's scientific project to show both why he adopted these tactics and how they contributed to the reception of his science and natural philosophy. It is important to realize, however, that despite Goethe's unconventional approach to studying nature, he was adamantly committed to the experimental method. Today, his science is often incorrectly labeled romantic, where romantic is considered an emotional or religious approach to nature rather than one that is based upon experiments or rigorous observations. An overview of his scientific corpus, however, demonstrates his deep commitment to the experimental method. One could even argue that the didactic volume of his *Theory of Colors* is Goethe's handbook on experimentation. In botany, his criticisms of Linnaeus had much to do with the procedures that Linnaeus used to collect data, while Goethe offered his own experiments as evidence to counter the theories of Wolff. Goethe's careful comparisons of various animal and human skulls led to his discovery of the intermaxillary bone, and he was an early champion of Luke Howard's (1772–1864) meteorological nomenclature because it facilitated the collection of data.

Once we have fully examined Goethe's scientific texts, we can better understand the revolution that he hoped to lead. In many ways, the philosophical principles underlying his science point the way to modernity, both in the humanities and in the sciences. At the same time, his scientific works also depart in significant ways from the premises of the various modern movements and remain critical of them. It is this interplay between his criticism of modern science, on the one hand, and his alternative philosophy, on the other, that comes to life in the study of each of these principles.

FAUST, GOETHE'S MOST FAMOUS literary character, perhaps is best known for his divided self. He laments that the two souls within his breast are tearing him apart. One desires spiritual knowledge and fulfillment, while the other craves physical satisfaction:

Two souls, alas! reside within my breast,
and each is eager for a separation:
in throes of coarse desire, one grips
the earth with all its senses;
the other struggles from the dust
to rise to high ancestral spheres.
(2: 30)

[Zwei Seelen wohnen, ach! in meiner Brust,
Die eine will sich von der andern trennen;
Die eine hält, in derber Liebeslust,
Sich an die Welt, mit klammernden Organen;
Die andre hebt gewaltsam sich vom Dust
Zu den Gefilden hoher Ahnen.]
(FA, I, 7: 57, 1112–17)

The discord between his souls, while bordering on the destructive (Faust quite seriously contemplates suicide), eventually spurs him on to a myriad of experiences and productive activity. Unlike the Faust in the traditional story or Marlowe's version of it, Goethe's Faust is not condemned to hell for his nearly boundless desire to experience all aspects of life. Rather, he is rewarded for his divided desires, and his two souls are ultimately reconciled and ascend into heaven.

The polar conflict within Faust's person is not by any means the traditional opposition between body and mind. It is instead a rejection of both the classical and the Christian emphasis on spirit. Faust finds only frustration in his single-minded pursuit of theoretical knowledge—whether of philosophy, law, medicine, or (especially) theology (FA, I, 7:

33, 354–64). Even his scientific endeavors within his laboratory leave him feeling imprisoned. He yearns to experience life outside (FA, 1, 7: 34, 386–409). Faust's condition is also an implicit rejection of the Cartesian duality of body and soul. Goethe explicitly spiritualizes the physical aspect of Faust's yearning by placing it within a soul and not limiting it to the body. In addition, after Faust's death, the physical part of his self initially ascends into heaven along with his spiritual one. Body impinges into the traditional realm of spirit, while spirit appears where one would expect to find body.

Polarity is Goethe's most basic universal principle. In emphasizing a polar principle, Goethe is partially reflecting and partially influencing a trend within the works of several of his contemporary scientists and philosophers. Kant (1724–1804), Schelling (1775–1854), Alexander von Humboldt (1769–1859), Carl Gustav Carus (1789–1869) and, of course, Hegel (1770–1831) all emphasized reciprocal, dialectical and/or oppositional relationships (Müller, 1–14). Of these thinkers, Goethe is most often closely linked to Schelling, whose early work he read and praised.[1] Like Goethe, Schelling placed polar forces at the center of his philosophy, turned to polarities to refute Newtonian science, and emphasized the creative functions of polarities. Goethe, however, differed from Schelling, as from many of the other Romantic scientists, in one very important respect: Goethe's principle arose primarily from physical observations and experience. Schelling was willing, and indeed believed it to be necessary, to base aspects of his polar philosophy outside of both experience and "the level of empirical science" (R. Stern, xv–xvi;).[2] In other words, for Schelling, polarities could be intuitive, transcendental, and prior to the experience and thus serve as a "higher" means of understanding physical phenomena (Ideen, 208–209/Ideas, 171–72). According to Goethe, principles did not exist prior to experience, but instead evolved out of and were indeed a part of experience. Thus, although Goethe ultimately questions the reliability of the senses, this questioning arises from various experiences whether from specific experiments or from general observations of the outside world.

Throughout Goethe's scientific corpus, he argues that the polar principle is at the foundation of all creative acts because only oppositions bring about new products. His view of polar relationships is the basis for his rejection of the very possibility of objective science. This view is also closely related to his argument that a separation of the subject from the object or the mind from the body leads to a necessarily one-sided and incomplete view of the world (H. Adler, "Erkenntis," 278; Erpenbeck, "Wissenschaft," 1190). He therefore rejects Descartes's colorless and dead world of matter in motion as portrayed in his *Discourse* or *The World* (*Le Monde,* ca. 1633) because it fails to portray the world as it really is: a vital and dynamic place, full of color and life, a place, moreover, that can only be understood by taking its power over our minds and bodies into account. Similarly, whereas Newton suggests that nature may be comprehended by separating and analyzing its parts, Goethe emphasizes the necessity of examining how the various parts work together. According to Goethe's theory, colors do not arise when white light is broken into parts, but arise when light and dark interact with each other.

Polarity is Goethe's best-known and most-written-about scientific principle.[3] A reexamination of this principle, however, sheds new light on its creative function within Goethe's philosophical thought. Thus, whereas several recent scholars focus on polarity as a primarily destructive or negating force[4] and more traditional ones see it (especially in its manifestation of subject and object) as an endorsement of an unqualified trust in the senses and the sum of his natural philosophy,[5] a close study of polarity within Goethe's scientific works, particularly in relation to his other principles (chapters 2–4), paints a different and more complex picture of its function. Goethe's polarities are not antinomies or logical binaries,[6] but represent opposing forces that often work together in order to create. And while the subject-object relationship plays an important role within Goethe's natural philosophy, the relationship itself is made more complex by other polar forces. The subject is not a unified entity that interacts with an object; instead, the subject is itself split into a variety of polarities that represent different desires, including those between two souls and two different bodily halves. These desires, moreover, play an important role in Goethe's philosophy. Unless the subject possesses self-awareness of his or her own desires, they may unduly influence his or her perception of the object. For example, a person who believes in mechanistic laws will see nature's activities in this light. Even an individual body part, the eye, may act independently and engage in the world through polar means: it creates its own complementary colors as a result of its interactions with objects.

That the eye may create its own phenomena due to a polar interaction with the object leads one to question the reliability of the senses and ask whether the observed object really exists or whether the eye has created it. Because all of these polar entities engage simultaneously in unions and reunions in their quest for wholeness (not unlike the account of Aristophanes' circle creatures), it becomes quite difficult for the subject to determine what "exists." The senses do not act alone, but need to be evaluated alongside numerous other factors. Thus, Goethe's questioning of scientific objectivity rests in part on the *unreliability* of the senses.

Goethe's view of the senses, however, does not mean that he endorsed a postmodern account of the world. Although he questions the reliability of the senses, he does not do so to deny the existence of truth, but as an attempt to get closer to it. This reevaluation becomes the first step that any scientist must take, whether in questioning his or her own theoretical viewpoint or questioning the validity of perception itself.

Throughout his corpus, Goethe emphasizes the centrality of nature's creative urge—whether in his discussions of polarity generally, his accounts of the creation of the world, or his description of the production of color. In each of these cases, the creation processes are similar and related. His discussion of the production of color in *The Theory of Colors* mirrors the polar creativity of nature as depicted in his two creation accounts. The creative actions of God and the universe become analogous to the way in which our senses perceive phenomena. The eye, like God, requires a polar partner for creation, and the eye, like any random piece of matter floating in the universe, craves its opposite in order to have a whole.

The polar views that Goethe expounds in his various texts are directly related to his rejection of one of the main tenets of modern science. His exposition of polarity, from his basic description of the polar principle to his scientific application of this principle, leads him to challenge the presupposition of a neutral scientific observer and objective scientific knowledge. Because polar partners are intimately intertwined, it is impossible completely to distinguish the influence of the object upon the subject or vice versa. Only by studying the power that objects have upon us do we become aware of their creative powers as well as our own—powers that in fact influence our scientific results. As Wilkinson similarly argues, "reality" for Goethe "is neither in the subject nor in the object but in the activity-between" (137).[7]

The Democracy of Polar Pairs

Readers of Goethe's literary works have long been aware of the importance of polarity, whether within the structure (e.g., the two mirrored halves of his *Elective Affinities*) or thematic content (e.g., Faust's soul or Ottilie's tragic conflict between duty and passion) of his works. His scientific works, however, demonstrate the all-pervasiveness of this theory. They further make clear that polarity is not primarily a negative or destructive force, but a highly creative one. For example, in his *Theory of Colors*, he reports that any good natural observer will see the polar principle everywhere:

No matter how different their opinions, faithful observers of nature will agree that anything that appears and meets us as phenomenon necessarily implies an original division capable of union or an original unity capable of division, and that the phenomenon must present itself accordingly. To make two of what is one, to unify what is divided—this is the life of nature, the eternal systole and diastole, the eternal syncrisis and diacrisis, the inhaling and exhaling of the world in which we live, weave, and exist. (12: 274)

[Treue Beobachter der Natur, wenn sie auch sonst noch so verschieden denken, werden doch darin mit einander übereinkommen, daß alles, was erscheinen, was uns als ein Phänomen begegnen solle, müsse entweder eine ursprüngliche Entzweiung, die einer Vereinigung fähig ist, oder eine ursprüngliche Einheit, die zur Entzweiung gelangen könne, andeuten, und sich auf eine solche Weise darstellen. Das Geeinte zu entzweien, das Entzweite zu einigen, ist das Leben der Natur; dies ist die ewige Systole und Diastole, die ewige Synkrisis und Diakrisis, das Ein- und Ausatmen der Welt, in der wir leben, weben und sind.] (FA, 1, 23, pt. 1: 239, no. 739)

Goethe's language here anthropomorphizes the world along the lines of Platonic (*Timaeus*) and Neoplatonic mysticism. The entire world is a living organism that breathes and has a heartbeat. All visible phenomena follow a polar pulse of separation and reunion.

According to the Goethean view, every created phenomenon, in one way or another, owes its existence to and may be described by the principle of polarity. Within the universe of manifold change, polarities enable Goethe to describe that change while simultaneously accounting for pattern within it. Things come to be and pass away. Within their very life span, they mimic a pattern of permanence and change. Goethe's "Polarity" essay[8] praises the frugality or the economy (Sparsamkeit) of nature because, through polarity, it creates an infinite variety of phenomena and creations. He explains that polarity is present at the birth of all existence. All things must divide to come into being, but reunion is as much a part of this cycle as the initial separation: "Whatever appears in the world must divide if it is to appear at all. What has been divided seeks itself again, can return to itself and reunite" (12: 156) [Was in die Erscheinung tritt, muß sich trennen um nur zu erscheinen. Das Getrennte sucht sich wieder und es kann sich wieder finden und vereinigen] (FA, 1, 25: 143). Although polarities are comprised of two opposing forces, whose interaction may lead at times to the negation or destruction of one or both elements, polar pairs may also represent reconciliations (such as the magnet) that preclude the concept of negation and permit creativity.

Polarity is itself, according to Goethe, part of an opposed pair. Together with his other main philosophical principle, *Steigerung* (intensification), polarity symbolizes a kind of division. Steigerung represents spirit—polarity, matter. As Huber notes, polarity and Steigerung are the most universal of Goethe's natural principles in that they address the dynamic capacity of both animate and inanimate nature (863). Unlike the traditional, hierarchical relationship between body and soul, however, Goethe intertwines the two and does not place one above the other. In an analysis of the essay "Nature" (letter to Müller, 24 May 1828),[9] Goethe explains that, although he generally agrees with many of the sentiments within the piece, the concepts of polarity and Steigerung are missing from it. This is a notable omission, since he characterizes these two forces as the two main driving forces of nature:

> . . . the former [polarity] a property of matter insofar as we think of it as material, the latter [Steigerung] insofar as we think of it as spiritual. Polarity is a state of constant attraction and repulsion, while intensification is a state of ever-striving ascent. Since, however, matter can never exist and act without spirit, nor spirit without matter, matter is also capable of undergoing intensification, and spirit cannot be denied its attraction and repulsion. (12: 6)

> [. . . jene [Polarität] der Materie, insofern wir sie materiell, diese [Steigerung] ihr dagegen, insofern wir sie geistig denken, angehörig; jene ist in immerwährendem Anziehen und Abstoßen, diese in immerstrebendem Aufsteigen. Weil aber die

Materie nie ohne Geist, der Geist nie ohne Materie existiert und wirksam sein kann, so vermag auch die Materie sich zu steigern, so wie sichs der Geist nicht nehmen läßt anzuziehen und abzustoßen.] (FA, 1, 25: 81)

What begins as one of the most standard binary divisions ends as a dynamic, fluid motion of one polar side to the other. One polar force or thing cannot exist or function without the other. This moving relationship between spirit and matter is consistent with Goethe's portrayal of his other philosophical principles and his representation of nature as a whole. Pure spirit and pure matter are not isolated but interact with their opposites. Goethean polarities thus challenge the very premise of Cartesian science—the disembodied ego.

For Goethe, the dynamic interaction of polar pairs is not limited to mind and body, but is characteristic of polar relationships generally. In "Polarity," Goethe provides us with a list that allows us to see the scope of this principle:

Wir und die Gegenstände (We and the objects)
Licht und Finsternis (Light and darkness)
Leib und Seele (Body and soul)
Zwei Seelen (Two souls)
Geist und Materie (Spirit and matter)
Gott und die Welt (God and the World)
Gedanke und Ausdehnung (Thought and extension)
Ideales und Reales (Ideal and real)
Sinnlichkeit und Vernunft (Sensuality and reason)
Phantasie und Verstand (Fantasy and understanding)
Sein und Sehnsucht (Being and desire)

Zwei Körperhälften (Two halves of the body)
Rechts und Links (Right and left)
Atemholen (Breathing)
Physiche Erfahrung (Physical experience)
Magnet
(FA, 1, 25: 142–43)

This list shows the comprehensiveness of Goethean polarities. They are evident throughout disparate spheres, from a religious universe (God and the world) to a Cartesian one (thought and extension). Perhaps the most striking aspect of the list is that, along with traditional oppositions, it contains polarities within both spirit (two souls) and body (two halves of the body). In the process, the borders between spirit and matter are broken down. Polarities exist within spirit as well as within matter. Faust speaks about two souls because two conflicting desires, sensual and rational, move him. Unlike Plato or Aristotle, Goethe does not rank the rational and irrational parts as higher and lower parts of the same soul. And unlike the romantics, he does not privilege the subject over the object or emotions over the intellect. Indeed, it is Faust's unified soul that ascends into heaven. Similarly, in the example of the two body halves of "Polarity," it remains an open question whether Goethe is referring to the mirrored aspect of left and right, or whether the division cuts the body between the rational (head) and the desirous (genital organs), or both.

The positioning of the various elements of the pairs undermines the classic notion of the hierarchy of one element over the other. Within the list, at times the traditionally considered higher element comes first (the subject, God, and light), while at others the opposite is true (body, sensuality, and fantasy). Goethe's final two examples, breathing and the magnet, reinforce the equality of both sides. Inhaling and exhaling and the two poles of the magnet are equally important and mutually serve to define the activity and the object, respectively.

Goethe's views of polar interactions extended into his views on scientific methodology. In the essay "Analysis and Synthesis" ("Analyse und Synthese," ca. 1829),[10] he uses the image of inhaling and exhaling to describe the ideal scientific approach. Goethe does not in this essay elaborate or give examples that explain how a union of these opposed scientific methodologies may come about.[11] The discussion here, as elsewhere on this issue, remains vague. (And indeed, in practice, he so often condemned the analytical that he undermined the acceptance of his own project within the scientific community.) He begins by describing polar conflicts that may arise between two groups of people. He mentions the conflict between rational and emotional people and then focuses upon the conflict between synthetic and analytical approaches. He places himself in the middle of these conflicts. He then compares analysis and synthesis to breathing. Both should presuppose and demand the other, and both are equally necessary. He even speaks of science as a living organism, which can come to life—can breathe—only by simultaneously employing both concepts (FA, 1, 25: 84).

Goethe grounds his belief in the "whole" of nature upon his observations of it. Nature acts in ways that require both quantitative and qualitative approaches. And while Goethe does not give specific details as to how one is to combine the two approaches, he argues that a purely mathematical approach is an inadequate means of studying nature.[12] Part of the reason that Goethe is so focused upon the qualitative side is that he believed that Descartes and Newton had missed nature's most central aspect: its creative urge. Goethean entities, as opposed to Cartesian ones, possess a craving for wholeness that causes them to seek (or even to create) their polar opposites, whether the eye creates complementary colors, or Mephistopheles, the devilish character of *Faust,* admits he always accomplishes the good while wishing to do evil (*Faust I,* 1335). Before turning to Goethe's accounts of nature and perception within his scientific works, it is helpful first to turn to two nonscientific accounts of the origins of creation. In these accounts, the first taken from his autobiography, *Poetry and Truth (Dichtung und Wahrheit),*[13] and the second from the poem, "Reunion" ("Wiederfinden," 1815), one can begin to see how essential the polar principle is to Goethe's philosophy of natural creativity. In neither account is God the God of the Judeo-Christian tradition: he is not perfect or even omnipotent, is not involved in human affairs, and does not create the world ex nihilo. God exists in a meaningful way only to the extent that he partakes in a polar process of creation. In other words, God becomes a metaphor for nature's creative acts. These accounts, therefore, prepare the way, not for Goethe's metaphysics, but for his physics. The eye when it perceives color and the mind when it processes sensory information act exactly in the same manner as God does in the act of creation.

Polarity and Creation Myths

Goethe begins the creation myth in his autobiography by discussing God as a single, unified entity. Such an entity can create nothing. God eventually splits into three parts, and these parts then represent the complete and whole of God and closes the "circle of divinity" (4: 262) [Hiermit war jedoch der Kreis der Gottheit geschlossen] (FA, 1, 14: 382).

Divine completeness—the triad of divinity—arises only through a kind of division. However, as a whole entity or closed circle, God is again incapable of further creation. The creation of a corruptible world by a perfect God is, of course, an old dilemma. The God of Genesis creates the world ex nihilo and thereby avoids the paradox of a perfect God creating a corruptible world. Goethe's godhead creates the world quite differently. He makes a polar opposite to himself, Lucifer, in order to create the world through a contrary act, a contradiction (ein Widerspruch).

The relationship between Lucifer and God resembles the one between polarity and Steigerung. It begins with the appearance of separation and opposition, but eventually becomes an interrelated one. Lucifer represents all that is material and concentrated in creation, while the godhead symbolizes all that is spiritual and expansive. Lucifer, however, soon tries to revolt from the godhead and concentrate power within himself. Yet, as the sole presence of God could not bring about creation, so too Lucifer also fails to bring about creation. The whole universe, as a consequence, heads for collapse: "the universe of course lacked its better half. For while it possessed everything that is gained by concentration, it lacked everything that can be accomplished by expansion" (4: 262) [so fehlte freilich dieser Schöpfung die bessere Hälfte: denn alles was durch Konzentration gewonnen wird, besaß sie, aber es fehlte ihr alles was durch Expansion allein bewirkt werden kann] (FA, 1, 14: 383). Neither God as the spiritual agent nor Lucifer as the material one may create without their polar opposite. Lucifer is not an entirely evil agent, nor is God a superior force when examined in the context of creativity.[14] Both forces are necessary for creation to occur.[15]

The godhead decides to contribute the polar force of expansion to Lucifer's contraction. This polar pair becomes the pulse (der eigentliche Puls) of both life and creation. Only after the godhead's contribution does the creation myth of the Judeo-Christian tradition occur: "This is the epoch when the thing we know as light emerged and when the process began that we customarily designate with the word 'creation'" (4: 262–63) [Dieses ist die Epoche, wo dasjenige hervortrat, was wir als Licht kennen, und wo dasjenige begann, was wir mit dem Worte Schöpfung zu bezeichnen pflegen] (FA, 1, 14: 384). Each subsequent act of creation continues to mirror this archetypal and polar one. The creation of light mirrors the creation of Lucifer, the fall of human beings from grace mirrors Lucifer's fall, and sexual reproduction mirrors the role of both Lucifer and the godhead in creation.

Although human beings are similar to everything throughout creation, Goethe also describes them as experiencing the world in a more heightened way. They are "the most perfect and the most imperfect, the happiest and the unhappiest of creatures" (4: 263) [das Vollkommenste und Unvollkommenste, das glücklichste und unglücklichste Geschöpf] (FA, 1, 14: 384). While human beings can never remain in a godlike position because of their corruptible nature, their striving toward perfection and knowledge holds out the promise that progress is possible. Bennett, in his account of this passage, addresses only the "tragic aspect" of creation and characterizes the pulse in our own lives as a "helpless oscillation between headlong blundering and deadly inertia, between destructive ignorance and ingenious knowledge, between being blind and being dazzled" (95). The human task according to Goethe's philosophy, however, is not a Sisyphean one, where one is forced to repeat the same task over and over again. Rather, Goethe's notion of intensification (chapter 2) describes how the very polar efforts, once combined with intensification, change both the individual and the task so that the individual possesses the potential to grow and develop. Moreover, as the passage demonstrates, the negative and unhappy side of Lucifer's being is just as instrumental in creation and in creating as the more positive and godly side.

Goethe's creation myth demonstrates two central ideas within his natural philosophy. First, it establishes that human beings are fully integrated with nature (although they are also one of its most intensified forms) because they participate in the same pulse of creation as the rest of nature or, for that matter, the same pulse of creation that existed between God and Lucifer. This pulse of activity is not simply reflected in the moral sphere, but in the sensory one as well. Within *The Theory of Colors,* the senses act according to the same principles as God (as is discussed below), so that we imitate the first principle of creation every time we look at the world. Second, as a direct consequence of the first, the creation myth underscores why Goethe instructs his readers to study both the material and the "spiritual" or vitalistic elements of nature.

Both are equally present at the birth of all things, and both are also equally present in every sensory act.

Goethe's poem "Reunion" ("Wiederfinden") goes even further in defining God primarily as the polar principle of creation. Although God instigates all creation, he completely withdraws by the end of the poem. He is able to do so because polarity continues the process of creation without his direct participation. The poem, however, also adds a new element to the discussion. The creation myth in Goethe's autobiography stresses that creativity requires the equal participation of both polar sides. In "Reunion," the desire of polar halves for their opposites comes into play. The mutual coexistence of entities does not insure their interaction. Nature's various parts, both animate and inanimate, must first crave to be whole in order to unite with their opposites. Goethe's nature comes alive only through an impulse that cannot be quantified. This impulse, moreover, is at the heart of why in his scientific essays he is so critical of scientists who fail to take the qualitative aspects of nature into account. He argues that they not only overly mechanize and regulate a willful force, but they also fail to recognize their own desirous relationship with their objects of study. This craving of one side for another ultimately undermines the very possibility of objective scientific research.

The title of the poem, "Reunion," refers to a reunion both of opposed lovers (Süßer, lieber Widerpart) and opposites generally. The main body of the poem refers to past time, the time of separation. God creates the world in the poem through division. Following the Judeo-Christian tradition, creation begins with God's speech, but unlike this tradition, the process before and after God's utterance is described in terms of a painful childbirth:

> When the world lay in the depths
> Utmost of God's eternal breast,
> Creative with delight sublime
> He willed a moment to be first;
> As he spoke the word "Become!",
> An anguished [painful] "Ah!" rang out, the All
> Exploded [powerful throes] with a motion vast
> Into being actual.
> (*1: 213)

[Als die Welt im tiefsten Grunde
Lag an Gottes ew'ger Brust,
Ordnet' er die erste Stunde
Mit erhabner Schöpfungslust,
Und er sprach das Wort: 'Es werde!'
Da erklang ein schmerzlich Ach!
Als das All, mit Machtgebärde
In die Wirklichkeiten brach.]
(FA, I, 2: 490)

Once God has decreed creation, it proceeds outside of him through polar divisions:[16]

Light was opened wide: the dark
Withdrew from it with diffidence;
Separation, scattering
Clove then apart the elements.
(1: 215)

[Auf tat sich das Licht! So trennte
Scheu sich Finsternis von ihm,
Und sogleich die Elemente
Scheidend auseinander fliehn.]
(FA, I, 2: 490)

The act of division, however, highlights the imperfection of the creator and of creation itself. It also draws our attention to God's creative essence. Goethe's imperfect God of the poem, who feels pain at creation, now feels loneliness. The Adam of Genesis is lonely because no animal is an appropriate companion for him. Adam requires a rational companion. The God of Goethe's poem, however, does not need a rational companion, but an activity to mirror his own. The simple existence of matter does not guarantee the further development of the universe. The God of the poem is lonely because his creation lacks all aesthetic qualities. It is silent and colorless. The world he has created is at this point similar to Descartes's created universe in *The World*[17] (or to a Lucretian one if the atoms did not have the ability spontaneously to swerve). There is motion, but it has no purposeful direction. There is activity, but it is not moved by desire. This God has no companionship because he is a creative being, but nothing creative exists. Nature is lacking its essential vitalism. Opposites exist in the universe that he has created, but they are incapable of reconciling. They feel no desire for each other. Yet a reunion of the opposing elements is necessary if creation is to be made possible.

To solve the problem of dead matter in space, God creates the red light of dawn. Dawn is a reconciliation of opposites. It reconciles both light and darkness and day and night. (In the next chapter, I further discuss how red, the color of dawn, also represents a heightened reconciliation of opposites in *The Theory of Colors*.) Once this reconciling force is created, the whole of nature becomes alive with desire, "And everything that fell apart / Now could fall in love again" (1: 215) [Und nun konnte wieder lieben, / Was erst auseinander fiel] (FA, I, 2: 491). The will of nature is its enlivening quality and becomes the basis for all natural phenomena. It causes inanimate matter to attract and repel and organic matter to unite and separate. Love, like dawn, serves as a linking force between two opposites and enables their reunion through mutual desire:

> Beings, if they do belong,
> Each seeks the other in its place;
> Sight and feeling hurtle them
> Back to life that measureless.
> Grasp or snatch, no matter how,
> Take hold they must, if they're to be:
> Allah's work for now is done,
> Creators of the world are we.
> (1: 215)

> [Und mit eiligem Bestreben
> Sucht sich was sich angehört,
> Und zu ungemess'nem Leben
> Ist Gefühl und Blick gekehrt:
> Sei's Ergreifen, sei es Raffen,
> Wenn es nur sich faßt und hält!
> Allah braucht nicht mehr zu schaffen,
> Wir erschaffen seine Welt.]
> (FA, I, 2: 491)

As soon as the created objects have an inclination toward one another, all kinds of unions, reunions, and new creations are possible. The universe, as we know it, begins to take shape. The inclination of nature's parts for one another is therefore central for any understanding of Goethe's philosophy of nature. All of nature's parts, even inorganic, possess a drive that leads to creation. For Goethe, the modern conception of the world as matter and motion is truly a dead one. Without the desire of one entity for another, no interaction or life, color, sound will occur.

In the poem, God, inanimate and animate nature, and human beings all act in a similar way. Allah[18] need no longer create because all of nature now creates in his likeness and in his stead. He sets the basic principle of polarities and polar reconciliation into motion, and all creation follows this pattern. Within the context of the poem, the "we" (wir) need not be limited to human beings, but includes any entity in the universe that participates in the creation of new forms. The poem concludes by returning to the reunited lovers. Their reunion, as the initial one of light, darkness, and the elements, takes place at dawn and is also symbolic of polar reconciliation: "Together on the earth we stand / Exemplars both, in joy and pain" [Beide sind wir auf die Erde / Musterhaft in Freud und Qual]. These two lovers, the poet tells us, no longer need to worry about separation through creation: "And a second word: 'Become!' / Shall not tear us apart again" (1: 215) [Und ein zweites Wort: Es werde! / Trennt uns nicht zum zweitenmal] (FA, 1, 2: 491). Due to the presence of desire, creation no longer means permanent separation and loneliness, but leaves open the potential for new unions. For the readers of Goethe's scientific texts, this reunion also includes the reunion of mind and body and of subject and object.

Goethe's God becomes a metaphor for the process of polar creation. God merely initiates creation, enabling subsequent creation through the principle of polarity. He then departs and plays no more direct role within the universe. Whereas Aristotle's God is constantly engaged in thought, Goethe's is defined by a creative act. Goethe, within his scientific texts, similarly describes God's role in the natural world. In the essay "Toward a General Comparative Theory" ("Versuch einer allgemeinen Vergleichungslehre," 1794), he describes God as acting only indirectly in creation (FA, 1, 24: 213), whereas in his essay "Formative Drive" ("Bildungstrieb," 1820) he describes how the formative drive, or the active and creative drive of living organisms, is like a God (Gott), creator (Schöpfer), and preserver (Erhalter) "whom we are constrained to worship, honor, and praise" (12: 35) [welchen anzubeten, zu verehren und zu preisen wir auf alle Weise aufgefordert sind] (FA, 1, 24: 452). Goethe's God, whether in the poem or in these essays, is not then an active intellect or a patriarchal figure to human beings. He is defined not by omniscience or omnipresence, but as the instigator of polar creativity.

Goethe's emphasis on the polar impulse of creation, an impulse that enables sound, color, life, helps us to understand many of the more "scientific" concerns within his *Theory of Colors*. His complaints against Newton often center around Newton's inability to see the interrelationship and the creativity of polar partners. On a small scale, this meant that Newton had "falsely" explained color through refrangibility and the division of white light instead of accounting for it through the interaction of light and shadow. For Goethe, the dynamic interactions of such polarities as light and darkness, and not the angles of refraction and the separation of white light into different rays of color, become the key to understanding the phenomenon of color. On a large scale, by ignoring this basic part of nature, Newton, by claiming the exclusivity of his own analytical method, prevented scientists from studying the wonderful (herrlich) and the delightful (erfreulich) qualities of natural phenomena (FA, 1, 23, pt. 1: 1048).

The Theory of Colors

Goethe deemed his *Theory of Colors* the most significant of his entire corpus. Eckermann reports that Goethe would very late in his life repeatedly contend:

As for what I have done as a poet . . . I take no pride whatever in it. Excellent poets have lived at the same time with myself; poets more excellent have lived before me, and others will come after me. But that in my century I am the only person who knows the truth in the difficult science of colors—of that I say, I am not a little proud, and here I have a consciousness of a superiority to many. (*Conversations with Eckermann,* 19 February 1829)

[Auf Alles was ich als Poet geleistet habe . . . bilde ich mir gar nichts ein. Es haben treffliche Dichter mit mir gelebt, es lebten noch Trefflichere vor mir, und es werden ihrer nach mir sein. Daß ich aber in meinem Jahrhundert in der schwierigen Wissenschaft der Farbenlehre der Einzige bin, der das Rechte weiß, darauf tue ich mir etwas zugute, und ich habe daher ein Bewußtsein der Superiorität über Viele.] (FA, 2, 12: 320)[19]

Goethe's comments may seem surprising when viewed in the context of his efforts to refute the truth of Newton's theories on light and refrangibility.[20] And although these comments show a certain lack of self-perspective, they must also be taken within the greater context of Goethe's scientific goals. *The Theory of Colors* was important to Goethe because so much more was at stake for him than just refuting Newton.[21] In this work, he not only sets forth his principles of optics and color, but within those very principles, he reveals the basis of his worldview. For Goethe, Newton's "errors" were not simply about color, but about a way of thinking about the world. To use the terms of Goethe's creation myth, the emphasis that Newton places upon nature's material side (or those aspects of it that could be easily quantified) is a reflection of Lucifer's mistake.[22] In Goethe's mind, such a study of nature threatened to collapse science into too limited a field. For example, in an advertisement for his *Theory of Colors,* he compares the scientists from antiquity and the Middle Ages with Newton. He criticizes the narrowness of Newton's sphere of research because it excludes the more beautiful and phenomenological aspects of nature. Newton speaks of light rather than color, he breaks up light to explain color rather than analyzing the interplay of light and shade, and he never addresses the colors' influence upon the observer. In contrast, Goethe praises the older scientists, who were more free in their approach and not as limited as Newton because they examined the manifold phenomena of color and sought to gather a more complete (vollständige) view of it (FA, 1, 23, pt. 1: 1047–48). For Goethe's *Theory of Colors,* this means not only talking about the qualities of the colors, but their moral and aesthetic influence upon the subject. He thought that scientists ought to investigate not only the physiological explanations of color, but also such issues as why yellow is a cheerful color and red is often considered a majestic one. Such aspects of his color theory became of interest to several painters, but did not impress Goethe's scientific readers, who only became more convinced that he was a poet and artist, and not a scientist (Heisenberg, "Goethe" and "Die Goethe'sche"; Helmholtz, "Goethe's Presentiments" and "On Goethe's"). This tension represents the core problem of Goethe's science and his criticism of Newton. Goethe thought that Newton was too narrow a scientist because he only focused upon the objects and not upon how they might affect the observer. For modern science, however, this separation of the object from the subject is precisely the point of scientific investigation (Heisenberg, "Goethe," 64, and "Die Goethe'sche," 150).

The Theory of Colors is in some ways Goethe's magnum opus. It is his longest and most comprehensive work. It is comprised of three main parts: the Didactic Part in which he outlines his theory and provides comprehensive lists of scientific experiments, the Polemical Part in which he directly attacks Newton's theories and his person, and the Historical Part in which he provides a sourcebook or chronological anthology (beginning with the ancients and concluding with the late eighteenth century) of the most important writings on color. It was a very expensive book to publish as it also included hand-painted cards to accompany the experiments of the Didactic Part.

The most famous and influential part (and the topic of this chapter) was the Didactic Part. Unlike either of the other two parts, it has been translated into English several times—the first time in 1840 when it was translated by Sir Charles Eastlake, the later (1853) founder of the Royal Photographic Society. It was of great interest to the English Romantics and directly influenced J. M. W. Turner, who painted at least one painting, *Light and Colour (Goethe's Theory)—the Morning After the Deluge,* according to the principles outlined in the Didactic Part. This part is divided into six sections. It begins with the "Physiological Colors" or the colors that are the "property of the eye, dependent on effect and countereffect of the eye" (12: 165) [dem Auge angehören und auf einer Wirkung und Gegenwirkung desselben beruhen] (FA, 1, 23, pt. 1: 26). Each part increases in complexity (exploring colors created by prismatic experiments, chemical compounds, etc.), culminating in the sixth and last part, "Sensory-Moral Effect of Color" ("Sinnlich-sittliche Wirkung der Farbe"). This part is the most eclectic and explores such aspects of color as its ability to create mood, the propensity of certain peoples to favor particular colors, rules for painting, and allegorical uses of color.

The Didactic Part of *The Theory of Colors* does not stress laws but creative formations. At the center of this web of reunion and creation are the eyes: our physical eye, the organ that perceives and creates color, as well as our spiritual, mind's eye, which also perceives color, but in its "moral effects," and which also creates, but in a more spiritual context.

While most of the text examines the physical eye, Goethe—at times tacitly, while at others overtly—also addresses the activity of the spiritual eye and its efforts to gain in knowledge. In other words, knowledge is based upon perception gained through polar means, and knowledge itself follows a kind of polar pattern. Almost everyone agrees that Goethe's use of the polar principle within his *Theory of Colors* did not lead to major scientific developments. However, even Goethe's harshest critics admire his keen sense of observation, and it is this sense that is central to his philosophy: Goethe is not wrong when he discusses the polar process of the physical eye, and it is this process that he then uses to form his epistemology. As Schrimpf has already argued in his discussion of subjectivity, the philosophical perspective is much more relevant than the scientific one (135).[23]

In the preface to his *Theory of Colors,* Goethe asks us to consider light in the same way we would consider the character of a human being: by studying its actions and passions (Taten und Leiden).[24] Goethe is assuming we already know something about how our "spiritual" eye works through our past experience with human character. His book on colors is meant to train our physical and spiritual eyes to work together to observe nature generally and colors in particular. He talks about colors as the actions and passions of light because he believes that light, like human beings, cannot be understood by being reduced to mathematical principles. Instead, he wants his readers to study the actions and reactions of particular colors within a context: how they react against different backgrounds, how they change when they are mixed with other colors, how they make the observer feel, etc. Goethe's first step and main intention (Hauptabsicht) of the whole work is to familiarize us with the language of nature (Natursprache), a language he hopes will "facilitate among the friends of nature the communication of higher views" (my translation) [. . . und so die Mitteilung höherer Anschauungen unter den Freunden der Natur zu erleichtern] (FA, 1, 23, pt. 1: 13). The "grammar" of this language is the principle of polarity:

No matter how diverse, enigmatic and intricate this language often seems, its elements remain forever the same. With gentle weight and counterweight nature balances the scales as they swing. "Here and there," "up and down," "before and after," are dimensions that emerge in the course of this weighing and serve to make specific the phenomena we meet in space and time. (12: 158)

[So mannigfaltig, so verwickelt und unverständlich uns oft diese Sprache scheinen mag, so bleiben doch ihre Elemente immer dieselbigen. Mit leisem Gewicht und Gegengewicht wägt sich die Natur hin und her, und so entsteht ein Hüben und Drüben, ein Oben und Unten, ein Zuvor und Hernach, wodurch alle die Erscheinungen bedingt werden, die uns im Raum und in der Zeit entgegentreten.] (FA, 1, 23, pt. 1: 13)

Polarity, as Goethe goes on to explain, is part both of movements and structures and can be perceived in a variety of ways:

as simple attraction and repulsion, as the waxing and waning of light, as the motion of air, as vibration of solid bodies, as oxidation and reduction. All these, however, have the effect of dividing or uniting, of setting existence in motion and lending support to some form of life. (12: 158)

[bald als ein einfaches Abstoßen und Anziehen, bald als ein aufblickendes und verschwindendes Licht, als Bewegung der Luft, als Erschütterung des Körpers, als Säurung und Entsäurung; jedoch immer als verbindend oder trennend, das Dasein bewegend und irgend eine Art von Leben befördernd.] (FA, 1, 23, pt. 1: 13)

As in the creation myths, polarities are at the source of creation and can be found everywhere. He further comments that many others have noted the principle of polarity in nature and have given it a variety of terms, including plus and minus, action and passion, and masculine and feminine (FA, 1, 23, pt. 1: 13). These terms are also a part of the language of nature and function as symbols and metaphors.

Learning the language of nature is not easy. One cannot ever hope to master it, Goethe admonishes, simply by reading the text of his work. Rather, he instructs his readers that they must become diligent students and practice the exercises in the book so that they may become fluent and experience the language directly in nature. Like a language textbook, Goethe's *Theory of Colors* begins with the most basic and simple concepts and then slowly builds upon them. It begins with the concepts of black and white and ends with the discussion of the moral effects of colors.

It is important to note how much emphasis Goethe places upon the visual aspects of the language of nature. This language speaks as much to our eyes as to our rational minds. He compares, for example, his own attempts to explicate nature in scientific experiments to a dramatic play:

A good play is only half present in the written text. The greater portion of it draws on the glitter of the stage, the personality of the actor, the power of his voice, the distinctiveness of his gestures, even the intelligence and favorable mood of the audience. This applies all the more to a work on natural phenomena. If the reader is to enjoy and make use of it, he must actually have nature before him, either in fact or in the activity of the imagination. It should be as though the writer were speaking in person; his text should be a directly visible demonstration of the phenomena as they occur naturally or artificially. Then all of his commentary, clarification and interpretation will succeed in creating the effect of life. (12: 162)

[Denn wie ein gutes Theaterstück eigentlich kaum zur Hälfte zu Papier gebracht werden kann, vielmehr der größere Teil desselben dem Glanz der Bühne, der Persönlichkeit des Schauspielers, der Kraft seiner Stimme, der Eigentümlichkeit seiner Bewegungen, ja dem Geiste und der guten Laune des Zuschauers anheim gegeben bleibt; so ist es noch viel mehr der Fall mit einem Buche, das von natürlichen Erscheinungen handelt. Wenn es genossen, wenn as genutzt werden soll, so muß dem Leser die Natur entweder wirklich oder in lebhafter Phantasie gegenwärtig sein. Denn eigentlich sollte der Schreibende sprechen, und seinen Zuhörern die Phänomene, teils wie sie uns ungesucht entgegenkommen, teils wie sich durch absichtliche Vorrichtungen nach Zweck und Willen dargestellt werden können, als Text erst anschaulich machen; alsdann würde jedes Erläutern, Erklären, Auslegen einer lebendigen Wirkung nicht ermangeln.] (FA, 1, 23, pt. 1: 18–19)

Instead of explaining polarities through language or formulas, Goethe requires that his readers conduct the experiments so that they can witness the polar examples for themselves. The language of the text alone is not as rich as when it is combined with visual and emotional experiences. One is more likely to see additional interpretive possibilities when the text is made visual and one is also able to "feel" the influences of particular colors oneself. Also, if his readers conduct the experiments themselves, they are much less likely to be confused or led astray by the particular wording of passages. Goethe's goal, however, is not simply to enable his readers to understand nature better: he wants his readers to understand their own reactions to specific phenomena as well. By comparing color experiments to a drama, Goethe emphasizes that these experiments are relevant to understanding aspects about our emotions and intellectual processes. For Goethe, scientific research was not only about discovering things about the objects but also about discovering things about ourselves. He therefore devotes an entire section to the book outlining the effects and impressions that particular colors have upon us, as well as investigating their possible symbolic significance.

Goethe stresses in this passage the multiplicity of perspectives and influences with any scientific experiment. Scientific "truth" does not lie with an "objective" observer. Nor is "truth" a single event. Rather, one needs to be mindful of the relationship between the polar opposites of subject and object. In watching a play, its meaning is mediated through several stages. One has not only the author's written word (or the scientist's theory), but also the actions of the actors (or the objects of study) that may be totally independent from the text. These actors might have their own wills apart from the text that need to be taken into account in analyzing the final interpretation of the play. In addition, each performance (or each particular experiment) will have its own dynamic, including such intangibles as the mood and the receptivity of the audience. By approaching science only from the perspective of the subject/author, scientists may miss other interpretive possibilities and thereby brush over some of the complexities and beauty of the object under study.

Within the main body of *The Theory of Colors,* Goethe's discussion of nature's language centers around the polar phenomena of light and darkness together with their closely related colors, yellow (the color most closely affined with light) and blue (the color most closely affined with darkness). He begins by addressing the eye's reactions to light and darkness, black and white, and complementary colors and concludes by arguing that visible polar phenomena have corresponding moral phenomena. He believes most fervently that his experiments with light and shade prove that Newton is wrong. Where Newton argues that colors arise from the breaking up and the refracting of white light, Goethe argues that all colors are procreated (Erzeugung [FA, 1, 23, pt. 1: 27]) or produced by the marriage of light and shade (FA, 1, 23, pt. 1: 26).[25] Blue arises when darkness or shade predominates within the union of light and shade, and yellow arises when light predominates over darkness. Moreover, Goethe's very method within *The Theory of Colors* reflects the process of generating color. In the process of creating his own scientific text, he places some facts "in the light" and while leaving others "in the shadows" [manches in's Licht, manches in Schatten setzen werde] (FA, 1, 23, pt. 1: 17). Like the creation myth, where all actions mirror their process of creation, Goethe's scientific method mirrors the polarity of nature itself. He consciously models his own scientific method upon the polar principle at the basis of all of creativity. In so doing, he attempts to describe the whole of nature, its material and its more inner, spiritual sides.

Goethe's arguments about light and shade are at the center of his refutation of Newton's Opticks. In prismatic experiments, Goethe believes refraction cannot explain the production of color, but that color arises only where light is bounded by darkness, shadows, or non-light (no. 691).[26] No colors will form, according to Goethe's theory, without the interplay between light and shadow. For example, he replicates the first experiment of Newton's Opticks and places red and blue squares against a black background.[27] In this experiment, Newton proposes that "[l]ights which differ in Colour, differ also in Degrees of Refrangibility" (20). When these squares are viewed through a prism, Newton found that, "if the refracting Angle of the Prism be turned upwards, so that the Paper may seem to be lifted upwards by Refraction, its blue half will be lifted higher by the Refraction than its red half. But if the refracting Angle or the Prism be turned downward, so that the Paper may seem to be carried lower by Refraction, its blue half will be carried something lower thereby than its red half" (21).

Goethe, in replicating Newton's experiment (nos. 258–84), also places blue and red squares against a white background. He does so to prove that the interactions of the colors with the particular backgrounds will directly influence the results: their actions and passions will change whenever the background changes. Although, in the end, Goethe's experiment does not challenge Newton's theory, he is correct in his observations of the phenomena. Black and white backgrounds will produce diametrically opposed results. Although the blue square appears to "be lifted higher" on the black background, it will appear lower on the white. Similarly, in those instances when the red square appears higher on the black background, it will appear lower on the white. The color of the background will change the direction of the displacement. According to Goethe, the color context or the interaction of the colors with the different backgrounds will determine the final results.

Goethe went even further with this set of experiments in his attempts to show that the phenomena described by Newton had "nothing" to do with refraction, but were the "result" of the interactions of light and shade. Accordingly, the blue and red squares only appear to move up and down due to their relationship with the polar elements of light and shade. If one looks closely at the colored squares through a prism, one notices that both are bordered with strips of color—a phenomenon not addressed by Newton. For instance, on the black background one may see a yellow/red border on top of both the red and blue squares, while on the bottom of the squares a blue/violet border appears. These border colors arise in Goethe's interpretation because, in this instance, red and blue act as light (they are both lighter than black), and when the prism causes the squares to be displaced across the black background, the subsequent interaction of light and darkness will produce the border colors. (Goethe establishes this point through several experiments using prisms and black and white surfaces and objects [nos. 197–257].) Red, therefore, only appears to move higher because the yellow/red color border blends together with the red of the square. The result: the red square looks longer on that side. Conversely, on the other side, where the blue border has appeared, the blue square looks longer. Goethe explains:

The effect of the homogeneous and heterogenous borders described here in detail is so powerful and strange that at first glance a casual observer will have the impression that the two squares have been pushed out of their horizontal alignment and displaced in opposite directions, the red square pushed upward, the blue one downward. But this apparent effect will not deceive those who know how to observe methodically and connect or develop experiments systematically. (no. 267)

[Die Wirkung der homogenen und heterogenen Ränder, wie ich sie gegenwärtig genau beschrieben habe, ist so mächtig und so sonderbar, daß einem flüchtigen Beschauer beim ersten Anblicke die beiden Vierecke aus ihrer wechselseitig horizontalen Lage geschoben und im entgegengesetzten Sinne verrückt scheinen, das Rote hinaufwärts, das Blaue herabwärts. Doch niemand, der in einer gewissen Folge zu beobachten, Versuche an einander zu knüpfen, aus einander herzuleiten versteht, wird sich von einer solchen Scheinwirkung täuschen lassen.] (FA, 1, 23, pt. 1: 106–7)

Again, although one may certainly question the applicability of these observations in overthrowing Newtonian theory, Goethe's account of the phenomenon is quite exact and correct.[28] For Goethe, the critical component of the experiment is the interaction of light and darkness. Opposite backgrounds (such as the white and the black) will cause opposing color results. So, too, will the interaction between light and darkness create color phenomena (the colored borders across the colored squares). In Goethe's mind, Newton's conclusion that different colors have different angles of refraction arises at the very least out of self-deception (Selbstbetrug), if not out of intentional deception

(Täuschung [no. 268]), dishonesty (Unredlichkeit) and "hocus-pocus" (FA, 1, 23, pt. 1: 315, no. 45). Goethe's harsh language against Newton probably did more damage to Goethe's reputation as a scientist than did his unconventional and at times even mystical approach. In essence, Goethe accuses Newton of railroading the reader: first Newton explains the conclusion before he has his reader conduct the experiment, making it more difficult to look for alternative explanations; second, Newton fails to mention the border colors that are not accounted for within his theory and indeed in Goethe's mind served to contradict it; and, third, Newton takes a very simple experiment and makes it as difficult as possible to give himself credibility and confuse the reader (FA, 1, 23, pt. 1: 310–16, nos. 34–46).

The very ordering of Goethe's experiments is designed to counter Newton's methodology. Newton's very first experiment requires the use of an instrument: the prism. Goethe does not address prismatic experiments in his *Theory of Colors* until after he has examined a much more basic aspect of perception: the workings of the eye. Goethe does not believe that one needs prisms to demonstrate the role of polarities in creating color, but argues that one sees colors produced by polar interactions with the naked eye all of the time. By applying Goethe's metaphor of the theater, his entire work may be read as a play. The various actions and interactions of light and darkness and their representative colors of yellow and blue form a plot. The actions and passions of light in *The Theory of Colors* constitute a drama, in which yellow and blue play leading roles. Blue and yellow are opposites. Blue is closely affined with darkness, or the passive, negative pole; yellow is closely affined with light, or the active, positive pole. According to Goethe's theory, all colors in the color wheel have their source from one or both of these colors, and all colors are the result of mixing of light and shade. It is commonplace knowledge that the result of physically mixing yellow and blue is green. What is not, however, intuitive is Goethe's treatment of pure red (purpur). Pure red, according to Goethe, arises either out of the intensification of yellow or blue, or the meeting of the two. Yellow intensifies into orange, which then darkens to approach red. Blue similarly deepens into violet, which then heightens to approach red.

The claim that yellow and purple, and blue and orange, are complementary colors is fairly well known and also easily established. Goethe tells us to "hold a small piece of lively colored paper or silk against a moderately illuminated white background, look steadily at the small colored surface for a time, and then remove it without shifting our gaze" (*12: 175) [Man halte ein kleines Stück lebhaft farbigen Papiers, oder seidnen Zeuges, vor eine mäßig erleuchtete weiße Tafel, schaue unverwandt auf die kleine farbige Fläche und hebe sie, ohne das Auge zu verrücken, nach einiger Zeit hinweg] (no. 49). He tells us the same results will occur if we leave the colored paper where it is but physically move the eye to gaze upon a white background. The result in both cases is that when the eye views a yellow piece of paper, it will create a violet aftershadow, or when viewing blue, it will create an orange one. This simple experiment is but one of many that focuses upon the eye's ability to create colors that do not exist in the object that it is viewing. Such experiments, therefore, become central to Goethe's arguments of polar creativity.

A closer look at Goethe's descriptions of these colors and the processes used to attain them demonstrates that he also endows them with spiritual significance. Although all of the colors are physical manifestations and concrete examples of both polarity and intensification, their spiritual meanings, to which he refers throughout the book, are not as apparent as their physical counterparts. He admits that these meanings, like language itself, are conventional, "for the significance of the emblem must be learned before its meaning is clear" (12: 296) [indem uns erst der Sinn des Zeichens überliefert werden muß, ehe wir wissen, was es bedeuten soll] (no. 917). Part of his goal in writing this book, however, is precisely to teach the "friends of nature" the meanings of such symbols. The double meaning of the phenomena (physical and spiritual) becomes evident throughout the book where various things, from organs to colors, take on different meanings. The importance of polarity throughout these meanings begins with the very first paragraphs of the text that address the eye.

Perception and the Physical and "Spiritual" Eyes

Although the formation of Goethe's colors demands some kind of mixing of the opposites of light and shade to produce color, this element is not the only polar aspect for the production of color. Throughout his *Theory of Colors,* Goethe discusses both the *physical eye* (the organ of sight) and the *spiritual eye* (the inner organ of intuition) (see figure 1).[29] Both of

Figure 1: *Goethe's Eye*. Woodcut modeled after a drawing by Goethe of his own eye. Photograph by Sigrid Geske. Reprinted by permission of Stiftung Weimarer Klassik/ Goethe-National Museum.

these eyes, according to Goethe, so crave wholeness that they create the polar opposites to the phenomena presented to them in order to "find" a whole. For Goethe, the eye represents in a microcosm the manner in which the human mind works: both strive for completeness, and both create opposites in order to do so. In other words, both the physical and spiritual eyes use polarity to create a sense of closure. And although this sense of closure is not static and may in the next moment lead to a new creation, for one moment it represents completeness and wholeness, whether a harmony of colors or philosophical fulfillment.

The manner in which Goethe discusses the eye in the introduction sets the tone for the rest of the work. Nature reveals itself through color to the eye. Just as Goethe's mystical account of creation in *Poetry and Truth* describes a triad that enables all further creation, so too does a triad of light, darkness, and color enable the creation of the entire visible world [Und so erbauen wir aus diesen Dreien die sichtbare Welt] (FA, 1, 23, pt. 1: 24). Moreover, in "Reunion," Goethe relates the creation of the world itself to this same triad. After Allah creates matter and light, no further creation is possible because no interaction takes place between light and darkness. It is a world that is desolate, chaotic, without sound and without color. Only after Allah creates dawn, i.e., color or the mixing and blending of light and darkness, does the world become alive and beautiful.

Goethe turns to this same "myth" in our daily lives when he discusses the eye. He writes that the eye is analogous to light and is indeed created by light in its own image: "From among the lesser ancillary organs of the animals, light has called forth one organ to become its like, and thus the eye is formed by the light and for the light" (12: 164) [Aus gleichgültigen tierischen Hülfsorganen ruft sich das Licht ein Organ hervor, das seines Gleichen werde] (FA, 1, 23, pt. 1: 24). Light, like the biblical God, creates in its own image. The eye continues the process of creation begun by the light, just as the "we" of the "Reunion" continue the process of creation begun by Allah. For example, within the main body of the text, Goethe characterizes the afterimage of a complementary color as springing out (entspringt) of an image that belongs to the eye (dem Auge angehört [no. 49]). The eye, then, is powerful because it, like light or God, may create out of itself and in its own image. To reinforce this point, Goethe quotes a poem based upon pre-Socratic, Ionic philosophy:

> Were the eye not of the sun,
> How could we behold the light?
> If God's might and ours were not as one,
> How could His work enchant our sight?
> (12: 164)

> [Wär' nicht das Auge sonnenhaft,
> Wie könnten wir das Licht erblicken?
> Lebt' nicht in uns des Gottes eigne Kraft,
> Wie könnt' uns Göttliches entzücken?]
> (FA, 1, 23, pt. 1: 24)

One of the poem's points is that like can only recognize like, and therefore the eye must be similar in some way to light.[30] Goethe, however, goes even further than admitting the similarity between light and the eye (and hence by analogy the similarity between human beings and God), and posits a relationship of equality. What begins as a proportional relationship light/eye as God/human beings (both light and God create in their own image, which then makes them recognizable to their created objects) becomes a relationship of equality:

None will dispute a direct relationship between light and the eye, but it is more difficult to think of the two as being simultaneously one and the same. We may clarify this by stating that the eye has within it a latent form of light which becomes active at the slightest stimulus from within or without. We can evoke dazzling inner images in the dark through the power of our imagination. In dreaming, we can see objects as though in the clear light of day. When awake, we can perceive the slightest impression of light from without, and we can even find that when the eye is struck a burst of light and color is seen. (12: 164)

[Jene unmittelbare Verwandtschaft des Lichtes und des Auges wird niemand leugnen, aber sich beide zugleich als eins und dasselbe zu denken, hat mehr Schwierigkeit. Indessen wird es faßlicher, wenn man behauptet, im Auge wohne ein ruhendes Licht, das bei der mindesten Veranlassung von innen oder von außen erregt werde. Wir können in der Finsternis durch Forderungen der Einbildungskraft uns die hellsten Bilder hervorrufen. Im Traume erscheinen uns die Gegenstände wie am vollen Tage. Im wachenden Zustande wird uns die leiseste äußere Lichteinwirkung bemerkbar; ja wenn das Organ einen mechanischen Anstoß erleidet, so springen Licht und Farben hervor.] (FA, I, 23, pt. I: 24–25)

The organ of the eye has the potential, which it often actualizes, to create just as light created the eye. The main difference between the light and the eye (or between God and human beings) is the question of potentiality: light and God are perpetually active, whereas the eye and human beings are only intermittently so. The eye and human beings, during their moments of creativity, are lightlike and godlike, respectively. According to Goethe's explanation, the potential power can be made actual through either thought (imaginative) processes or through physical ones (direct contact). This same theme of creation—of bringing forth out of oneself—is present in Goethe's autobiography and in "Reunion." The "Ionic" poem that Goethe cites, as well as his explanation of it, combine all three accounts of creation: the eye, light, and God are all mentioned, and all three create by combining opposites. The last two lines of the poem may be interpreted as arguing that if the light = the eye, then, by the analogy that is presented within the poem, God = human beings.

Goethe admits the difficulty (Schwierigkeit) of recognizing this new formulation, but emphasizes the eye's ability to call forth light (thereby creating in its own image), as well as darkness and color. The eye creates out of itself, reacting to the sensible world, the whole triad responsible for the creation of the visible world. The eye begins to look like the original unified image of God before he splits up into the triad. In other

words, just as the autobiography and "Reunion" argue that human beings possess, in so far as they utilize their polar powers, the creative
power of God, this scientific work contends that the eye participates in
the same creative process. The similarities among the part (eye) and the
whole (human being) and the cosmic beginning (God's creation)
demonstrate the unity of all of nature. These similarities also reinforce
the centrality of creation in Goethe's philosophy—creation made possible only through the power of polarity. God, light, the eye, or even human beings use polarity to create the world, color, visible objects, art,
and even other human beings.

This same polar, creative process relating to the eye appears in concrete examples throughout *The Theory of Colors*. As Goethe begins to lay
down the foundations for the color theory itself, the eye continues to
play a crucial function:

For the eye, color is an elemental natural phenomenon. Like every phenomenon it
manifests itself in division and opposition, combination and union, intensification
[Erhöhung] and neutralization, infusion and diffusion, etc. and can best be observed and understood through these general principles of nature. (12: 164)

[Die Farbe sei ein elementares Naturphänomen für den Sinn des Auges, das sich,
wie die übrigen alle, durch Trennung und Gegensatz, durch Mischung und Vereinigung, durch Erhöhung und Neutralisation, durch Mitteilung und Verteilung
und so weiter manifestiert und unter diesen allgemeinen Naturformeln am besten
angeschaut und begriffen werden kann.] (FA, 1, 23, pt. 1: 25)

The eye, like Aristotle's concept of *nous,* is a thing whose potential is limitless because of its malleability and its ability to accept all kinds of
forms. Like the magnet, the eye may simultaneously contain two opposite states (nos. 13, 15), and the eye's creative powers stem from the very
fact that when presented with one phenomenon (such as blue) it creates
the opposite phenomenon (the complementary orange). Once again,
the eye is able to deal with polar opposites simultaneously (no. 33), while
it strives to create the whole. Goethe discusses the eye in terms of the vitality of all of creation:

We may infer that in this instance we have once again recognized the retina's great
vitality and the silent contradiction every living thing is moved to express when
presented with any specific state. Thus inhaling presupposes exhaling and vice
versa; each systole presupposes its diastole. Here, too, it is the eternal rule of life
which is asserting itself. *When offered something dark the eye demands something bright; it de-*

mands darkness when presented with brightness. Through this very fact it demonstrates its living qual-
ity, its right to take hold of an object, by bringing forth out of itself an element which is the opposite of
the object. (12: 173, emphasis added)

[Wir glauben hier abermals die große Regsamkeit der Netzhaut zu bemerken und
den stillen Widerspruch, den jedes Lebendige zu äußern gedrungen ist, wenn ihm
irgend ein bestimmter Zustand dargeboten wird. So setzt das Einatmen schon das
Ausatmen voraus und umgekehrt; so jede Systole ihre Diastole. Es ist die ewige
Formel des Lebens, die sich auch hier äußert. *Wie dem Auge das Dunkle geboten wird, so*
fordert es das Helle; es fordert Dunkel, wenn man ihm Hell entgegenbringt und zeigt eben dadurch
seine Lebendigkeit, sein Recht, das Objekt zu fassen, indem es etwas, das dem Objekt entgegengesetzt
ist, aus sich selbst hervorbringt.] (no. 38, emphasis added)

This passage summarizes many of the ideas already discussed. The eye
reacts to a physical phenomenon and is able to create in the mind another, contrary one. Its function is not simply one of perception, but also
creation. Its creative product, moreover, may confuse the observer and
should teach him or her to question the reliability of the senses. If the
eye is exposed too long to a particular color phenomenon, it will eventually create the opposite one to achieve a balance. The eye will create
white to counterbalance black, green to counterbalance red, etc. Where
Goethe's account of creation emphasizes the necessity of the presence of
polar opposites for any creation to take place, similarly here, in the very
act of sight, where images are "created," is polarity of central, functional,
and structural importance.[31] The eye, like God and light, creates in its
own image, "brings forth out of itself." Goethe places so much emphasis
upon the eye because he hopes that it will demonstrate the inherent creative power of nature's parts.

When Goethe discusses the eye, however, he does not limit it to the
physical organ, but intimates that the spiritual eye of human beings
functions in a similar way to the physical one.[32] The eye, as well as the
various colors, represents both physical and spiritual phenomena. In his
preface, Goethe describes how his *Theory of Colors* will lay out for the
thinking reader (dem denkenden Leser) how he or she can form a conception of the whole. He uses the same word (ein Ganzes) throughout
the text to describe the process of the eye creating polar opposites. The
spiritual eyes of the thinking reader will, by the end of the work, be able
to create a philosophical whole, just as the physical eye creates a physical
whole.[33] Within the text itself, Goethe refers to the physical *and* spiritual

eyes. In a passage where he emphasizes how he has avoided arbitrary symbolism or phrases, he explains how he has sought to put the phenomena themselves in front of our physical and spiritual eyes (vor den Augen des Leibes und des Geistes [no. 242]).[34]

Moreover, when Goethe discusses the eye's ability to form complete wholes by producing the opposite of the phenomenon presented to it, his language is filled with moral terms, so that he directly links the function of the physical eye with that of the spiritual one. In fact, he is quite clear that this property of the physical eye, the property that creates its opposite, is a means of achieving spiritual freedom. He specifically calls our full attention to the passage that discusses the eye's ability to create freedom for itself. After discussing the color circle with its complementary and polar colors, he announces, "Here we have arrived at an important point, one deserving our fullest attention" (12: 284) [denn wir stehen hier auf einem sehr wichtigen Punkt, der alle unsre Aufmerksamkeit verdient] (no. 811). In the next paragraph, he describes how viewing colors affects us somewhat pathologically because they exhibit an influence upon us and sweep us up into a variety of emotions, depending upon the particular color: "we felt lively and active, passive and anxious, lifted to exalted heights, or reduced to the mundane" (12: 284) [uns bald lebhaft und strebend, bald weich und sehnend, bald zum Edlen emporgehoben, bald zum Gemeinen herabgezogen fühlten] (no. 812). These emotions are a limitation (Beschränkung). The eye, demanding totality, however, frees itself from such limitation (of this imbalance toward one polar phenomenon) by bringing forth its opposite. It thereby forms a new whole and creates a new balance:

But the eye's inborn need for totality allows us to escape this limitation; it finds its freedom by creating the opposite of the color forced on it, thus producing a satisfying whole. (12: 284)

[so führt uns das Bedürfnis nach Totalität, welches unserm Organ eingeboren ist, aus dieser Beschränkung heraus; es setzt sich selbst in Freiheit, indem es den Gegensatz des ihm aufgedrungenen Einzelnen und somit eine befriedigende Ganzheit hervorbringt.] (no. 812)

While this freedom is fleeting—complementary colors quickly fade and new impressions are constantly placed in front of the eye that further demand the creation of new oppositions—for a brief moment the eye carves out freedom for itself. Goethe goes on in the next paragraph to generalize this principle of freedom. The eye is but one symbol of a harmony of opposites that liberates:

These truly harmonious opposites are simple but important as evidence that nature is inclined to set us free through totality, for here we are the direct beneficiaries of a natural phenomenon with esthetic implications. (12: 284)

[So einfach also diese eigentlich harmonischen Gegensätze sind, welche uns in dem engen Kreise gegeben werden, so wichtig ist der Wink, daß uns die Natur durch Totalität zur Freiheit heraufzuheben angelegt ist, und daß wir diesmal eine Naturerscheinung zum ästhetischen Gebrauch unmittelbar überliefert erhalten.] (no. 813)

Goethe deliberately draws our attention to the fact that our physical eyes are linked with our spiritual ones, and that freedom is achieved through both physical and spiritual wholeness. Once again, spiritual freedom, just like Goethe's conception of thought as a whole, finds its roots in observable physical phenomena. He therefore justifies including a discussion about intellectual and spiritual liberation within a scientific treatise. In his mind, practicing science includes the reactions of the observer to the phenomenon. He exhorts his readers to conduct these experiments for themselves and to observe the phenomena so closely that the idea of totality, of harmony, completely penetrates them and they can *feel* it in their spirit [uns endlich von der Idee dieser Harmonie völlig penetriert und sie uns im Geiste gegenwärtig fühlen] (no. 815).[35] Early in *The Theory of Colors* Goethe admonishes his readers to approach nature with freedom (Freiheit) in their search for the whole (FA, 1, 23, pt. 1: 14). Here, he tries to apply his teaching about color and freedom to the scientific method itself. Only by imitating nature's polarity within scientific experiments will the researcher find freedom and totality for the mind's eye.[36] It is for this reason that Goethe believes that the study of qualitative characteristics ought to be an integral part of science. Without such a study, researchers would not be aware of the imbalance in their own lives nor would they be able to "free" themselves from their particular perspectives. Goethe's metaphysical language is inappropriate within a scientific text, yet that is his point. His argument is that that is what is wrong with science: it ought to be able to include investigations outside of the quantitative. And although Goethe failed in bringing the scientific community around to his way of thinking, his reflections on the relationship between the subject and object led to a questioning of objectivity that seems quite modern.

Nearly twenty years after completing his *Theory of Colors,* Goethe still focused on the importance of the eye as a symbol of scientific and artistic endeavors. To make aspects of his work clear to Eckermann, he insisted that they read the work together, paragraph by paragraph. Only then would they have a solid object of entertainment (Gegenstand der Unterhaltung) before them and in this way could Eckermann come away with a whole teaching (die ganze Lehre). Once again, Goethe stresses that the eye itself contains colors: "[T]here is nothing without us that is not also within us; and that the eye, like the external world, has its colors" (1 February 1827) [es ist nichts außer uns, was nicht zugleich in uns wäre, und wie die äußere Welt ihre Farben hat, so hat sie auch das Auge] (FA, 2, 12: 228–29). The success of the scientific method itself depends upon understanding the properties of the eye. Goethe wanted scientists to distinguish between the properties (in this case of color) in the eye and those in the outside world. The eye, due to its creative abilities, may mislead the subject. The observer must question his or her perception in order to determine what really exists and what the eye has created. While the eye contains colors and is capable of producing them itself, Goethe carefully distinguishes between this creative act and the independence of the outside objects of colors:

Since a great point in this science is the decided separation of the objective from the subjective, I have properly begun with the colours belonging to the eye; that in all our perceptions we may accurately distinguish whether a colour really exists outside ourselves, or whether it is only a seeming colour produced by the eye itself. (1 February 1827)

[Da es nun bei dieser Wissenschaft ganz vorzüglich auf scharfe Sonderung des Objektiven vom Subjektiven ankommt, so habe ich billig mit den Farben, die dem Auge gehören, den Anfang gemacht, damit wir bei allen Wahrnehmungen immer wohl unterscheiden, ob die Farbe auch wirklich außer uns existiere, ob es eine bloße Scheinfarbe sei, die sich das Auge selbst erzeugt hat.] (FA, 2, 12: 229)

Goethe does not deny the existence of colors in the external world, but he requires that researchers first examine the color from all perspectives in order to determine how it has come to exist. Their further conversations about the eye's constant ability to change and create new colors leads to a general discussion of a law, "which pervades all nature and on which all life and all the joy of life depend" [das durch die ganze Natur geht und worauf alles Leben und alle Freude des Lebens beruhet] (FA, 2, 12: 229). This "law of required change" (Gesetz des geforderten Wechsels) appears not only in the eye and in all of the senses, but "also with our higher spiritual nature. Because the eye is so eminent a sense that this law of required change is so striking and especially clear with respect to colours" [auch mit unserem höheren geistigen Wesen; aber weil das Auge ein so vorzüglicher Sinn ist, so tritt dieses Gesetz der geforderten Wechsels so auffallend bei den Farben hervor und wird uns bei ihnen so vor allen deutlich bewußt] (FA, 2, 12: 229). Goethe links this law (which explains the eye's ability to create the contrary of the phenomenon presented to it) to the use of major and minor scales, good style, theatrical performances, Shakespearean tragedies, and ultimately even to reflections about Greek drama. Good artists know that to create a rich work of art they must balance one emotion or style against another. Too much of one characteristic will make the piece boring or unpleasant. Although still warning Eckermann to see some of the uses of the law only by analogy, Goethe still stresses how far-reaching they can be: "You see how all things are connected with each other, and how a law respecting the colour theory can lead to an inquiry into Greek tragedy" [Aber Sie sehen, wie alles aneinander hängt, und wie sogar ein Gesetz der Farbenlehre auf eine Untersuchung der griechischen Tragödie führen kann] (FA, 2, 12: 230).

Goethe explicitly explains the importance of polar balance for aesthetics in several critical essays. Artists are judged by their abilities to incorporate conflicting elements within their work. For example, he praises Shakespeare for his ability to reconcile ancient and modern artistic tendencies. In his essay, "Shakespeare Once Again" ("Shakespeare und kein Ende," 1815), Goethe contrasts the two opposing characteristics of ancient and modern literature:

Antik (ancient),	Modern.
Naiv,	Sentimental.
Heidnisch (pagan),	Christlich (Christian).
Heldenhaft (heroic),	Romantisch (romantic).
Real,	Ideal.
Notwendigkeit (necessity),	Freiheit (freedom).
Sollen (duty),	Wollen (desire).

(FA, 1, 19: 641–42)

Goethe praises Shakespeare precisely because he is able to tie together (verknüpfen) the two opposing worlds within his literature: a central factor of Shakespeare's greatness lies in his ability to reconcile opposing elements (FA, 1, 19: 644). If one wants to learn anything from Shakespeare, Goethe argues, one must first study how he is able to combine the conflicting elements of ancient and modern drama (FA, 1, 19: 643–44).

Throughout this section, I have emphasized the polar power of the eye to create its own objects in reaction to phenomena presented to it. Goethe rejects the possibility of objectivity in large part due to this reciprocal relationship of the eye and the object of its perception. As the examples of the eye have demonstrated, one first needs to consider whether the object one is examining really exists or has been created by the subject through a reaction to another object. Moreover, it is impossible to be fully objective because the subject never is completely free of influences from the object or vice versa.

The Subject and Object and the Argument against Hierarchy

In Goethe's natural philosophy, neither the spirit nor the subject is placed in privileged positions above the body or the object. Within his discussions on polarity, he time and again argues for a joint rule between polar pairs and not for a set hierarchy. First, his observations of nature taught him that polar pairs exist and function only in reference to each other. One cannot have inhaling without exhaling, the systole without the diastole, or even reason without passion. Second, and most importantly, he traces all creative activity back to polar interactions, where both polar sides are necessary for any act of creation to occur. These forces explain the formation of everything from colors to human character and account for phenomena as diverse as aftershadows to aesthetic works of art. Goethean polarities thus challenge the subject-oriented, reason-based philosophical tradition. His principle of polarity rejects by its very definition the hierarchy (ascribed to Plato and Aristotle) of soul over body, Cartesian dualism, and the Kantian emphasis on the subject because such views deny the interrelatedness and the constant dynamic motion of the phenomena. For Goethe, a full understanding of the subject-object relationship helps one discover two main mistakes of modern science: ignoring the power of the object (as already discussed) and failing to recognize the subject's own desires within any relationship. Goethe admits that it is very natural for us to consider ourselves to be in a hierarchical position over the objects of our study. It therefore becomes the task of the scientist, philosopher, or even artist to try to see the reciprocal relationship between the two. Like the fluid relationship of polarity and Steigerung, or body and spirit, so too do subject and object exist in a mutually dependent relationship. This relationship, in the end, explains why Goethe rejects the very possibility of a neutral, scientific observer who may claim complete objectivity in his or her own work. As Fink notes in his discussion of Goethe's essay on experimentation, the "experiment, in Goethe's view, serves as a vehicle for transmitting the subject and object of investigation. That is, Goethe is not subscribing to the popular notion that the scientist is dispassionate and detached from his experiment. His view assumes passion and enthusiasm for the object under investigation" (38). Similarly, Sepper writes that the leading question of Goethe's essay is "whether there can be any way of ensuring that the subject, the researcher, can discover what is intrinsic to the object of study—whether the researcher can help the object provide its own meaning rather than simply reinforce the opinion of the researcher" (*Contra,* 66).

One of Goethe's main pieces of advice to scientists is to stop considering their observations of nature within a vacuum. In several works, including his *Theory of Colors,* "Maxims and Reflections," and "Experiment as Mediator between Subject and Object," he warns against the human propensity to see oneself as a sole creator of phenomena, precisely because he attributes powers of creation to the object as well. Even an individual organ, the eye, contains powers of creation that may confuse an observer. Goethe's conception of the universe is so creative that objects may be just as creative as subjects. *The Theory of Colors* shows how literally he views the creative abilities of objects: at times they pathologically impede our ability to use our senses, at times they morally influence our mood, and at times they cause us to create new objects that have no physical existence except as they relate to the initial objects. Thus, whereas Helmholtz criticizes Goethe's science as relying too much upon the reliability of the senses ("On Goethe's," 16), Goethe's scientific works actually warn scientists against trusting the senses too much. One must constantly analyze one's perceptions in order to distinguish between information about an object and the senses' own creative powers. Scientists wishing to study an object need to gain an ironic distance from their senses and themselves generally. Goethe calls it a "very natural way" (ganz natürliche Art) to view objects according to our own desires and passions. He also warns of the consequences: "it can lead to a thousand errors—often the source of humiliation and bitterness in our life" (12: 11) [und doch ist der Mensch dabei tausend Irrtümern ausgesetzt, die ihn oft beschämen und ihm das Leben verbittern] (FA, I, 25: 26). It leads to bitterness because, if researchers become wedded to their own particular way of observing nature, they may become frustrated in their attempts to study it. Viewing nature solely from the perspective of the subject tells only one part of the story about nature and the way it functions. As a remedy, he suggests that researchers ask themselves whether they or the object is exerting an influence and to try to view objects in their own right (FA, I, 25: 26–27; FA, I, 13: 50, no. 1.320). This approach to the subject-object relationship led Heisenberg to dismiss Goethe as a scientist. To Heisenberg, science "represents the attempt to describe the world to the extent that it is independent of our thought and action" ("Goethe," 67, and "Die Goethe'sche," 153). Yet, for Goethe, such a statement would epitomize what is wrong with modern science: a belief in the possibility of purely objective knowledge.

Goethe views the relationship between subject and object in light of a power struggle. To focus only upon the power of the subject is to miss the power that objects may have. Arthur Schopenhauer, who had occasion to discuss scientific issues with Goethe, reports the following encounter:

But this Goethe . . . was so thoroughly a realist, that it did not make sense to him that objects as such exist only in so far as they are conceived by the cognizant subject. "What," said he once to me, looking at me with his Jupiter-like eyes, "Light exists only in so far as you see it? No, you would not exist, if the light did not see you!"

[Aber dieser Goethe . . . war so ganz *Realist*, daß es ihm durchaus nicht zu Sinne wollte, daß die *Objekte* als solche nur da seien, insofern sie von dem erkennenden Subjekt *vorgestellt* werden. Was, sagte er mir einst, mit seinen Jupitersaugen mich anblickend, das Licht sollte nur da sein, insofern Sie es sehen? Nein, *Sie* wären nicht da, wenn das Licht Sie nicht sähe.] (*Goethes Gespräche*, 2: 937, original emphasis)

For Goethe, the subject must be conscious of the power of the object if he or she wishes to understand the reciprocal relationship of subject and object. Both exercise power over the other, and for the subject to ignore this power is a kind of "arrogance" for which Goethe upbraided Schopenhauer. Only those natural observers who admit that the objects themselves exist and have power over them will be able to progress in their studies because only then will they fully understand the dynamic way in which nature operates.

In a short essay, "The Enterprise Justified" ("Das Unternehmen wird entschuldigt," 1817), Goethe describes the relationship between the central polarity of subject and object. He rejects as prejudiced the Kantian notion of the primacy of the subject. Examining the world only through the perspective of the subject limits one's understanding both of the world and of oneself. Goethe instead postulates a more interdependent, flowing relationship as more representative of knowing the world, where the power of both subjects and objects is recognized:

When in the exercise of his powers of observation man undertakes to confront the world of nature, he will at first experience a tremendous compulsion to bring what he finds there under his control. Before long, however, these objects thrust themselves upon him with such force that he, in turn, must feel the obligation to acknowledge their power and pay homage to their effects. When this mutual interaction becomes evident he will make a discovery which, in a double sense, is limitless; among the objects he will find many different forms of existence and modes of change, a variety of relationships livingly interwoven; in himself, on the other hand, a potential for infinite growth through constant adaptation of his sensibilities and judgment to new ways of acquiring knowledge and responding with action. (12: 61)

[Wenn der zur lebhaften Beobachtung aufgeforderte Mensch mit der Natur einen Kampf zu bestehen anfängt, so fühlt er zuerst einen ungeheuern Trieb, die Gegenstände sich zu unterwerfen. Es dauert aber nicht lange, so dringen sie dergestalt gewaltig auf ihn ein, daß er wohl fühlt wie sehr er Ursache hat auch ihre Macht anzuerkennen und ihre Einwirkung zu verehren. Kaum überzeugt er sich von diesem wechselseitigen Einfluß, so wird er ein doppelt Unendliches gewahr, an den Gegenständen die Mannigfaltigkeit des Seins und Werdens und der sich lebendig durchkreuzenden Verhältnisse, an sich selbst aber die Möglichkeit einer unendlichen Ausbildung, indem er seine Empfänglichkeit sowohl als sein Urteil immer zu neuen Formen des Aufnehmens und Gegenwirkens geschickt macht.] (FA, I, 24: 389)

While the subject begins its relationship with the object, believing itself to be the one in control, very soon thereafter the object begins to exert its control over the subject. Although Bennett is correct that Goethe often warns against the human trait "to become so absorbed" in viewing our world that "we forget our true creative role" (44), he overstates the case when he concludes that objects exist only in human thought (45). Instead, the relationship is one of mutual dependence between subject and object. The eye represents one example of such a relationship. The eye begins by viewing one object of color, only to create another. The initial color limits the eye, but then the eye's ability to create a complementary color frees it by creating a whole. Goethe generalizes the reciprocal relationship between subject and object and asks scientists to consider the possible influence of the objects upon themselves. Acknowledging the power of the objects will have two consequences. First, the observer will expand his or her understanding of objects. They will not be seen as simple, dead objects with certain quantitative qualities, but as things that are alive with activity. Just as the world was animated by the desire of one polar half for the other in "Reunion," scientific research will come alive by recognizing the power of the object to draw the subject toward it. Second, the observer will expand his or her own powers of observation. By considering the possible powers of the object, whether upon mood, aesthetic sensibilities, or physical reactions, Goethe wishes to open science up to more qualitative study. This very flexibility, he promises, will lead to personal growth. Scientists will expand their ability to gather knowledge.

Goethe further discusses the philosophical importance of acknowledging the power of the objects in his essay "Significant Help Given by an Ingenious Turn of Phrase" ("Bedeutende Fördernis durch ein einziges geistreiches Wort," 1823). Here he shows how important objects are to his ability to think at all. He claims that material objects are never separated from his thoughts but are actually part of his thought processes. Even thoughts have material qualities. He agrees with the assessment that his "ability to think is objectively active" [Denkvermögen *gegenständlich* tätig sei] (FA, 1, 24: 595, original emphasis) and explains that the thinking subject cannot ultimately be separate from the material object.[37] He lambastes the Delphic dictum of "Know thyself" [erkenne dich selbst] as "a ruse of secretly allied priests" [eine List geheim verbündeter Priester] because it leads to a false kind of introspection. The Socratic life of introspection is to be rejected because it emphasizes spirit

at the expense of the outside world.[38] Only with the melding of both opposites—whether spirit and matter or subject and object—is an ideal re-
alized and knowledge obtained: "The human being knows himself only
insofar as he knows the world; he perceives the world only in himself,
and himself only in the world. Every new object, clearly seen, opens up a
new organ of perception in us" (12: 39) [Der Mensch kennt nur sich
selbst, insofern er die Welt kennt, die er nur in sich und sich nur in ihr
gewahr wird. Jeder neue Gegenstand, wohl beschaut, schließt ein neues
Organ in uns auf] (FA, 1, 24: 595–96).

For Goethe, knowledge of self does not arise solely from rational introspection but arises when one attempts to understand every aspect of
the relationship to the physical world. His concept of experience (Erfahrung) was thus very closely linked to knowledge of all kinds (H.
Adler, "Erfahrung," 272–73, and "Erkenntnis" 277–79). Self-knowledge
comes about through a study of the outside world because we, as individuals, are always engaged in a polar relationship with the objects
around us. To understand ourselves, we must also understand something
about these objects and our relationship to them. Subjective introspection (or, for that matter, the Cartesian cogito) leads to an intellectual
dead end.

Goethe more clearly outlines what this thought process is like within
his "Polarity" essay. He begins with the observation that two demands
(Forderungen) arise when one observes natural appearances (Naturerscheinungen): one wants to gain complete knowledge of the phenomena and one wants to gain possession of them through reflection. These
two goals are achieved through polar processes. One first observes the
objects, then grasps (fassen) them, and then finally reproduces them in
one's own mind (im Geiste). Not only do objects clearly have independent existences outside of the mind, but one gains mastery over them
(eine gewisse Herrschaft erlangen) only by first recognizing their exteri-
ority and materiality before reproducing them spiritually. Once again,
thought processes are inseparable from the objects, and knowledge is
achieved through a reciprocal relationship between subject and object.

Perception and knowledge for Goethe rely equally upon ourselves
and upon the object itself. Nor does he limit this material thinking only
to his scientific endeavors, but claims the same process occurs within
some of his poetry (FA, 1, 24: 596). He discusses his inclination to write

occasional poetry as a species of objectively active thinking. Often these poems are the result of a very specific interaction with a particular object. The object has such an irresistible (unwiderstehlich) influence upon him that its presence may then be found again within the images of his poetry. After discussing his poetry, he emphasizes again that the same process occurred in his botanical, osteological, and geological studies. For example, the fortuitous discovery of a broken and partially buried skull led to his theory of skull formation. Upon viewing the skull, it became immediately apparent to him that parts of the skull were really modified forms of vertebrae. For Goethe, objects possess a power within themselves that draw us toward them and enable us to progress in artistry and knowledge once we recognize their influence upon us. It is for this reason that he is suspicious of experiments that occur in the laboratory or ones that seek to understand entities by studying their parts to the exclusion of the whole. Both approaches presuppose that the object as a whole does not influence the subject. In writing up his own scientific discoveries, Goethe included what many consider to be extraneous information, as when, in the case of the aforementioned example of the skull, he discusses the inspirational occurrences behind it. He does so because, in his conception of science, it is important for the reader to know what his particular state of mind was in order to evaluate the discovery.

It is not, however, only the objects that exert an influence upon the final results. Goethe argued that scientists always have a subjective agenda when studying natural objects, and they must acknowledge their own desires for any particular object. For any one to claim objectivity is to ignore the inherently subjective endeavor of any scientific approach. In evaluating any scientific theory, one must not only be aware of the object's power, but also of the subject's desires within that theory: "In observations of nature large and small, I have uninterruptedly asked the question: Is it the object or is it you that is here expressing itself? And in this sense I also looked at my predecessors and my colleagues" (my translation) [Bei Betrachtung der Natur im Großen wie im Kleinen hab' ich unausgesetzt die Frage gestellt: Ist es der Gegenstand oder bist du es, der sich hier ausspricht? Und in diesem Sinne betrachtete ich auch Vorgänger und Mitarbeiter] (FA, I, 13: 50, no. 1.320).[39] Because a scientist's character and propensities inevitably influence his or her work,

Goethe, as a means of evaluating scientific research, tries to see how much the individual scientists have imposed themselves upon their observations of nature. He therefore was quite interested in studying the biographies of scientists alongside their discoveries. For instance, he especially requested biographical information about Luke Howard, the meteorologist, so that he could better understand his research (FA, 1, 25: 235). His interest in Descartes's *Discourse on Method* also led him to read Baillet's biography of Descartes (Diary, 17 and 26 February and 6, 7, and 23 March 1809), just as his investigation of Wolff's theories led him to study and eventually publish biographical information on him (FA, 1, 24: 426–28). For Goethe, understanding something about the life and character of the person could very well shed light upon that person's work.

At times, Goethe relies too heavily upon biographical information in his assessment of scientific work. For example, his judgment of the moral worth (der sittliche Wert) of Luke Howard's scientific works relies upon Howard's *autobiographical* account of himself (FA, 1, 25: 235). In other words, whereas Goethe warns that one should be wary of accepting objective scientific facts, he takes the most subjective of statements—a person's account of himself—at face value. Goethe, however, felt so strongly that an intimate connection existed between one's scientific product and one's personal life that he himself provided a wealth of autobiographical background to his own works. He published many explanatory essays alongside his scientific ones that provide the context of his scientific works, the personal impetus behind their composition, and the history of their reception (e.g., FA, 1, 24: 399–438, 732–52).

Goethe warns that one can never fully recognize the object's power because "it is evident that we theorize every time we look carefully at the world" [und so kann man sagen, daß wir schon bei jedem aufmerksamen Blick in die Welt theoretisieren] (FA, 1, 23: 14). Scientists' studies are always influenced by their particular backgrounds. The kinds of questions that they ask will influence the way in which they approach nature. Similarly, in "Empirical Observation and Science" ("Das reine Phänomen," 1798), he argues,

For the observer never sees the pure phenomenon with his own eyes; rather, much depends on his mood, the state of his senses, the light, air weather, the physical object, how it is handled, and a thousand other points. (12: 25)

[Denn da der Beobachter nie das reine Phänomen mit Augen sieht, sondern vieles von seiner Geistesstimmung, von der Stimmung des Organs im Augenblick, von Licht, Luft, Witterung, Körpern, Behandlung und tausend andern Umständen abhängt.] (FA, 1, 25: 125)

Subjects always have some particular agenda, influence, or purpose in how they look at the world. This, however, is Goethe's point. The subject can never be totally objective. The best that the subject can do is to become aware of his or her own desires toward an object as a means of beginning to understand the power of the subject-object relationship. In observing the world, Goethean scientists are to approach the world "with consciousness, with self-knowledge, in a free way, and (if I may venture to put it so) with irony" (12: 159) [mit Bewußtsein, mit Selbstkenntnis, mit Freiheit, und um uns eines gewagten Wort zu bedienen, mit Ironie zu tun] (FA, 1, 23: 14). The first step to freeing oneself from the limitations of one's perspective is to become aware of what that perspective is. Self-knowledge and irony imply the ability to step outside of oneself and gain a new perspective. Unlike Socratic self-knowledge, however, one is to use this ability to return to the world and study it. There is no such thing as completely objective knowledge because each subject will approach nature differently and hence possibly obtain different results.

To Goethe, the process of incorporating the power of the object within our study of it was not only a scientific proposition, but a philosophical one. He argues that viewing the world and one's place within it as a vital, interlocking relationship between polarities such as subject and object leads to the possibility of endless development (die Möglichkeit einer unendlichen Ausbildung), to a high level of enjoyment (einen hohen Genuß), and towards the path of perfection (dem schönen Lauf zur Vollendung) (FA, 1, 24: 389). In other words, a correct understanding of polarities and our own relationship to them leads to happiness and perfection. The scientist's quest becomes related to the Faustian one. Like Faust, who attempts to reconcile the two opposing forces of reason and passion within his own self, scientists should strive to reunite the divided entities of subject and object.

Conclusion

Throughout Goethe's scientific works, his polarities offer an integrated view of the world, which combines such diverse elements as the

organic and the inorganic, the human and the nonhuman, and reason
and passion. Their function is primarily a creative one: they are at the
structural basis of all creative acts from the creation of the universe to
theatrical performances. Polar interactions further illustrate a nature
that is alive due to its dynamic desire to form a whole.

Goethe criticizes Newtonian science throughout *The Theory of Colors*
because he believes that it offers an incomplete description of light,
color, and the workings of nature as a whole. Newton failed to address
the polar interactions that Goethe argued were at the base of all color
formations. By failing to address the role of polarities, Newton had
robbed the discussion of the central characteristic of optics: color.[40] In
contrast to Newton, Goethe argues that one cannot discuss optics without discussing the objects of vision themselves as well as their surround-
ing contexts. He places all aspects of color—from color formation to the
moral effects of color—at the center of his discussion. Goethe further
accuses Newton of unnecessarily complicating the discussion by his use
of intricate and involved experiments, where simpler ones could even
better explain the phenomena (FA, 1, 13: 90, no. 1.613; 407–408, no.
6.38.6). Goethe instead advocates turning to polarities in place of complicated experiments or formulas as a model for scientific inquiry generally (FA, 1, 25: 83–86). Within *The Theory of Colors,* he argues that the
basic principles of all color formation involve visible polarities that dy-
namically interact and that are not subject to complex formulas. James
Gleick, in *Chaos: Making a New Science,* writes of how scientists of this century have once again turned to Goethe's scientific approach (163–65,
197).[41] Feigenbaum, among others, was fascinated by Goethe's science
precisely because Goethe investigated dynamic principles instead of
fixed, natural laws, and because he attempted to study them in their en-
tirety rather than dissecting the objects of his experiments.

Goethean polarities present a nonhierarchical relationship between
opposites that emphasize creativity rather than rationality. Goethe, like
many postmoderns, does not embrace reason as the central point of departure for his work, and this is why he rejects Newtonian science. At the
same time, however, Goethe does not reject reason entirely, but sees it as
part of a polar pair. He is also very critical, as several scholars have no-
ticed, of the Romantics for their overemphasis on subjectivity.[42] Whereas
the Platonic or Aristotelian soul is arranged according to a hierarchy of

reason over passion, soul over body, and masculine over feminine, the very principles behind Goethe's theory of polarity preclude such hierarchies, and embrace instead a kind of dynamic relationship of mind and body, reason and passion, masculine and feminine, good and evil, etc. Therefore, while important aspects of Goethe's works lie outside of the rationalistic tradition, they also are opposed to a current tendency to employ polarities as a means of negation and subversion. Although his philosophy, like deconstruction, employs contraries, his polarities may lead to reconciliation and creativity and not simply negation and contradiction.

This chapter has focused upon the complexities behind Goethe's polar principles and how polarity is an integral part of Goethe's notion of creativity. Goethean polarities, however, by themselves, do not explain the phenomena that Goethe finds to be the most interesting, complex, and highly developed. Such phenomena arise when polarities are coupled with Goethe's other main principle: Steigerung. The next chapter argues that polarities, once combined with Steigerung, illustrate another important aspect of Goethe's natural philosophy: an individual's drive toward progression and internal development. While polarities remain an essential part of the process, Steigerung becomes a necessary component if the individual is to break free of old patterns of development. Once again, individual growth and development are not limited to the human realm but find their roots in the organic and inorganic world.

2 STEIGERUNG AND THE DRIVE TOWARD COMPLEXITY

 GOETHE'S PRINCIPLE of *Steigerung* is often associated in his literary works with Faust's striving, culminating in his ascension into heaven, or with Ottilie's transformation in *Elective Affinities* from a simple child into a mystical icon. An inner drive in both characters propels them forward in the face of obstacles and ultimately distinguishes them from those around them. Try as he might, Mephistopheles is unable to lull Faust into a static moment of satisfaction, a moment that according to their pact would condemn Faust forever. Mephistopheles cannot understand Faust's character and therefore loses his claim upon his soul. Ottilie, too, remains an enigma. She "triumphs" over her body by renouncing food and drink, and, after her death, she appears to become a saint who can raise the dead and heal the sick. Although Eduard, her lover, tries to emulate her, he, in the end, recognizes a genius in her "martyrdom" that is inimitable.

Goethe's term, Steigerung, derives primarily from his scientific works. There, as in the literary texts, he explains the ability of particular entities to strive toward greater perfection or self-overcoming. He variously describes this process as a specialization, augmentation, greater articulation of form, increased freedom of movement, darkened color, or heightened spiritual awareness. Polarities represent nature's most simple creative desire: the desire for physical union. This desire is present everywhere. Steigerung, in contrast, represents a natural desire that is not necessarily evident in every natural being (although it is potentially present in both animate and inanimate beings): the desire to transcend matter. One may witness this desire in the attempts of some individuals to transform into more complex forms, to create something new, or to break free of pre-existing laws and patterns. Thus, in contrast to many scholars who equate nature and necessity in Goethe's literary works (e.g., Bielschowsky, 2: 383; Gilli, 553–65; Gundolf, 553; Milfull, 94; Reiss, 209; Staiger, 2: 495), this study argues that Goethe does not see nature as synonymous with necessity, but as an entity that may at times freely exert its will. And although the "willfulness" of Steigerung is certainly more evident in nature's non-regular patterns (chapters 3 and 4), even in nature's more regular processes, Steigerung represents nature's desire to tend toward complexity. Gross forms may metamorphose into specialized organs, or light and shade may intensify into beautiful shades of color. In each of these cases, Steigerung begins in the material, physical realm and then gradually, through an inner drive, leads the entity away from the material and toward the nonphysical. As Lucretius had argued many centuries before,[1] Goethe also contends that natural units (whether animate or inanimate) contain a drive that enables them to act unexpectedly. Unlike Lucretius, Goethe does not reduce this drive to a random motion or swerve. Rather, he attributes to natural entities a more directed drive toward complexity.

Goethe's principle of Steigerung has two main functions within his scientific texts. On the one hand, it describes and accounts for natural progression and specialization. These products of Steigerung are visible to the observer. An inner desire or drive in natural entities leads to the creation of highly specialized organs or the "purification" of "baser" fluids. Also, as I argue in chapter 3, Steigerung illustrates that while nature at times may follow regular principles or laws, it may also break free of those laws and patterns and lead to completely new organisms. The observable manifestations of Steigerung further play an important role in Goethe's scientific methodology. He argues that if scientists are to gain an understanding of nature, they have to imitate its processes in their study of it.

On the other hand, Steigerung has a pronounced metaphysical meaning for Goethe that makes it quite obvious that he is not practicing science in the modern sense and leads one seriously to question whether he is practicing science at all: Steigerung is as much about the progression of natural entities as it is about the scientist's spiritual reaction to them. Goethe recognizes that he is not practicing science in the usual sense of the word, but one of his stated goals is to expand the scope of science beyond Cartesian or Newtonian conceptions of it. Unlike traditional scientists, he believes that one ought to investigate all aspects of nature, even its more mystical manifestations. In addition, he claims scientific validity in examining nature's mystical influences because he believes that the physical manifestations of Steigerung are linked to more mystical processes. His science has the goal of incorporating feelings and intuition without departing from empiricism: it is not a search for nature or religion within the interior of the subject (as many of the Romantics tried to do), but an attempt to explain the subject's complex reaction to the world. He argues that observers will experience certain reactions upon witnessing the process of Steigerung that are directly related to that process. He claims, for example, that red, the most highly intensified color in his schema, is universally recognized as being a more elevated color.

This chapter examines both aspects of Goethean intensification within the context of nature's more regular processes: the development of a flowering plant and the production of "pure red" (purpur). Both examples demonstrate nature's inherent desire to overcome the physical as well as Goethe's metaphysical interpretation of this process. The next two chapters then continue the discussion to show how this desire or drive is also the force behind the creation of new and irregular forms through evolution and is responsible for the kinds of natural "insurrections" that lead to "abnormal" organic forms. In all of these cases, however, nature is constantly striving to recreate itself. By positing nature's ability to self-overcome, Goethe distances himself from the classical understanding of teleology. As in the case of Faust, the moment never stands still for nature. Once nature reaches an apparent end or a telos, it will again seek to overcome it and form something new. He therefore repeatedly rejects the idea of teleology in his scientific writings because he does not believe that the ends of nature are fixed. They, too, evolve and transform.

Nature's Inner Drive

Chapter 1 already touched upon the close interaction between polarity and Steigerung. Polarity is more closely associated with the material, Steigerung with the nonmaterial:

Polarity is a state of constant attraction and repulsion, while intensification is a state of ever-striving ascent. Since, however, matter can never exist and act without spirit, nor spirit without matter, matter is also capable of undergoing intensification, and spirit cannot be denied its attraction and repulsion. (12: 6)

[. . . jene [Polarität] ist in immerwährendem Anziehen und Abstoßen, diese [Steigerung] in immerstrebendem Aufsteigen. Weil aber die Materie nie ohne Geist, der Geist nie ohne Materie existiert und wirksam sein kann, so vermag auch die Materie sich zu steigern, so wie sichs der Geist nicht nehmen läßt anzuziehen und abzustoßen.] (FA, 1, 25: 81)

In the creation myth, polarity was personified in the characteristics of Lucifer, and Steigerung in those of the godhead. The materiality of the devil threatened to collapse the world until the godhead contributed its expansive drive. The nonmaterial, or the inward or spiritual side of creation, leads the material outside of itself so that it may expand and create rather than collapse and destroy. The relationship between the godhead and Lucifer (like that between polarity and Steigerung) is characterized as a pulse. Although these two forces are closely related and each operates in both the material and the nonmaterial realms, Goethe distinguishes their functions. The effects of polarity are physically visible. One can see, for example, the creation of complementary colors literally before one's eyes. Steigerung, due to its association with the "spiritual" side of nature, is more difficult to "see." It is the impelling drive in nature's processes and is most evident in moments of organic transformation. While Goethe emphasizes the necessity of viewing polarity and intensification as inseparable, he considers intensified things as more interesting than unintensified ones in three main respects. First, because intensified things presuppose polarities, they are by definition more complex. Second, intensified things are more capable of creating new forms. Polarities create through more predictable patterns. Intensification enables nature to transcend those patterns and create new patterns or forms for itself. Third, Steigerung represents the highest process of creativity not only in nature, but in human beings. Goethe believes that the creative processes for nature and human beings are similar but differ only in degree. By studying nature's inner drives, he argues, human beings can learn something about their own creative impulses.

Goethe most directly discusses creativity and the role of Steigerung in the essay "Polarity," in which he extols nature's creative powers and emphasizes that nature creates a myriad of life-forms and phenomena through the cooperation of polarity and Steigerung. Although polarities are at the center of all created products, they are not as free in their ability to create new things. The eye, for example, is forced to create a particular complementary color. Once Steigerung is integrated with polarity, however, it brings about many more creative possibilities than those arising from polar interactions or unions. For example, new colors arise out of more complicated processes than complementarity, and a plant strives during its lifetime to develop more complex forms. Where polarity creates new unions (light and shade procreating such colors as yellow and blue), Steigerung brings about the progression and development of the individual polar partners (enabling the genesis of "pure red" or the complex organs in a plant). Their subsequent unions, as a result, are also more complex:

Whatever appears in the world must divide if it is to appear at all. What has been divided seeks itself again, can return to itself and reunite. This happens in a lower sense when it merely intermingles with its opposite, combines with it; here the phenomenon is nullified or at least neutralized. However, the union may occur in a higher sense if what has been divided is first intensified; then in the union of the intensified halves it will produce a third thing, something new, higher, unexpected. (12: 156)

[Was in die Erscheinung tritt, muß sich trennen um nur zu erscheinen. Das Getrennte sucht sich wieder und es kann sich wieder finden und vereinigen; im niedern Sinne, indem es sich nur mit seinem Entgegengestellten vermischt, mit demselben zusammentritt, wobei die Erscheinung Null oder wenigstens gleichgültig wird. Die Vereinigung kann aber auch im höhern Sinne geschehen, indem das Getrennte sich zuerst steigert und, durch die Verbindung der gesteigerten Seiten ein Drittes, Neues, Höheres, Unerwartetes hervorbringt.] (FA, 1, 25: 143)

Goethe does not give concrete examples in this essay of what these higher and lower forms might be. What is clear in this essay, however, is that intensification is identified as the force behind the transformation of simple forms into complex and unique entities: "To do this [nature] uses the principle of life, with its inherent potential to work the simplest phenomenon and diversify it by intensification into the most infinite and varied forms" (*12: 156) [Sie [die Natur] bedient sich hierzu des Lebensprinzips, welches die Möglichkeit enthält, die einfachsten Anfänge der Erscheinungen durch Steigerung ins Unendliche und Unähnlichste zu vermannigfaltigen] (FA, 1, 25: 143).

In another essay, "The Formative Impulse [Drive]" ("Bildungstrieb," 1820), Goethe further describes what he means by organic striving. Here he directly links Steigerung or a formative drive or impulse with the metamorphosis of form. This short essay, which is a commentary on contemporary theories of development and procreation, tries to distinguish the material aspects of growth from nonmaterial ones. Goethe rejects Caspar Friedrich Wolff's explanation of development through *vis essentialis,* a generative power, as being untenable because it is based too much upon materiality and therefore offers a merely mechanical explanation for organic development. "[T]he question," therefore, "of what is to organize itself out of that substance remains a dark, incomprehensible point" (*12: 35) [und das was sich aus jener Materie organisieren soll bleibt uns ein dunkler unbegreiflicher Punkt] (FA, 1, 24: 451). For Goethe, matter alone cannot account for organic development, but one must look behind material appearances for the cause of organic change. Goethe therefore praises Blumenbach's *nisus formativus,* "an impulse, a surge of action which was supposed to cause the formation" (12: 35) [einen Trieb, eine heftige Tätigkeit, wodurch die Bildung bewirkt werden sollte] (FA, 1, 24: 451), because it looks outside of material causes to explain organic development. Goethe interprets Blumenbach's drive as a nonmaterial impulse within the organism that acts as a cause or an agent of development.[2] Goethe explains in greater detail what he believes this *nisus formativus* to be:

We can examine this assertion more quickly, easily, and perhaps more thoroughly, if we recognize that in considering a present object we must suppose an action prior to it, and in forming a concept of an action we must presume a suitable material for it to act upon. Finally, we must think of this action as always coexisting with the underlying material, the two forever present at one and the same time. Personified, this prodigy confronts us as a god, as a creator and sustainer, whom we are constrained to worship, honor, and praise. (12: 35)

[Betrachten wir das alles genauer, so hätten wir es kürzer, bequemer und vielleicht gründlicher, wenn wir eigenstünden daß wir, um das Vorhandene zu betrachten, eine vorhergegangene Tätigkeit zugeben müssen und daß, wenn wir uns eine Tätigkeit denken wollen, wir derselben ein schicklich Element unterlegen, worauf sie wirken konnte, und daß wir zuletzt diese Tätigkeit mit dieser Unterlage als immerfort zusammen bestehend und ewig gleichzeitig vorhanden denken müssen. Dieses Ungeheure personifiziert tritt uns als ein Gott entgegen, als Schöpfer und Erhalter, welchen anzubeten, zu verehren und zu preisen wir auf all Weise aufgefordert sind.] (FA, 1, 24: 451–52)

Goethe ultimately refers to the formative drive as analogous to the ancient concept of form (FA, 1, 24: 452). He speaks of the coexistence of form (Form) and matter (Stoff), much in the same way as he spoke of the coexistence of polarity and Steigerung or of the necessity of the presence of both the godhead and Lucifer in the creation myth. Both are always present and each presupposes the other, and both are necessary in any act of creation. Although Goethe's statement about the relationship of form and matter resembles Aristotle's treatment of causes, Goethe emphasizes a different aspect of this relationship. He argues that the relationship is primarily about the ability of the formative drive to bring about change. He postulates that one cannot see the freedom (Freiheit) of the formative drive (Bildungstrieb) without the concept of metamorphosis. The moments of transformation, as witnessed through a process of change, best illustrate nature's freedom to change and progress.

Goethe most clearly develops these ideas in his "Metamorphosis of Plants" where nature's striving transforms raw, undeveloped material into specialized forms that gradually leave their material states behind. Goethe, however, does not limit natural striving to the organic realm but, as his *Theory of Colors* illustrates, he sees an analogous drive in the inorganic realm as well. In both works, Steigerung also has consequences for scientists. The process of Steigerung serves as a model for the scientific essay itself: Goethe consciously models his own scientific texts upon the patterns of intensification that he notes in his observations of nature. His texts, in this respect, resemble Platonic dialogues or Aristotelian treatises, where the form and substance of the teachings are inextricably linked and reflected in each other.

The Metamorphosis of Plants

The Transcendence of Matter

In his scientific works, Goethe often describes the principle of Steigerung as a step-by-step progression. This description is most evident in his treatise on plant development, "The Metamorphosis of Plants" ("Versuch die Metamorphose der Pflanzen zu erklären," 1790).[3] This essay goes through the various steps in a plant's development and, in the process, one witnesses how the leaf begins in a crude material form only gradually to overcome this state and strive toward transcendence of matter.

A brief account of the development of this essay sheds some light on its importance for Goethe's thought. With the exception of his *Theory of*

Colors, "The Metamorphosis of Plants" is his most famous scientific work. It is well known, not so much because of its content (although most people who have heard of the essay know it is about the metamorphosis of the organ of the leaf), but because Goethe discusses its significance for his life in several works, including the *Italian Journey*, "Fortunate Encounter," and the poem "Metamorphosis of Plants." He variously discusses the essay's importance as signaling his growing interest in science, allowing an important friendship to form, demonstrating nature's creative drives, and serving as an analogy to the development of a perfect relationship between two lovers.[4] Goethe also wrote numerous essays explaining the history and development of this essay and tracing the history of its reception. The work's reception pained Goethe a great deal. First, he had a very difficult time in finding a publisher for the work, even though he had already established himself as a successful and well-known author.[5] Then, once it was published, he felt it was "met with a cold, almost hostile reception" (12: 67) [kalte, fast unfreundliche Begegnung] (FA, 1, 24: 402). He believed that his scientific endeavor was viewed with suspicion because of his reputation as a poet. And although the language of this essay is not as overtly mystical as in his later *Theory of Colors*, a close study of the "Metamorphosis" once again shows that it was not simply Goethe's reputation that served to alienate his scientific readers.

"The Metamorphosis of Plants" was Goethe's first published scientific work (1790) and was published two years after his monumental trip to Italy. He had traveled to Italy to break away from his bureaucratic duties in Germany in order to complete several works of literature, to fulfill his lifelong dream of studying classical forms and art, and to further his scientific research. He also hoped that he would find the elusive *Urpflanze* or archetypal plant. He believed this Urpflanze would be an actual plant that would represent all other plant forms. The farther south he traveled, the more convinced he became that he would find this form. While in Sicily, he found that his scientific thoughts took precedence over his artistic ones. He found it impossible to finish his *Odyssey*-inspired play, *Nausikaa*. Each time he attempted to meditate upon his poetic dreams (meine dichterischen Träume), another spirit (Gespenst) would seize him. He could not stop thinking about the archetypal plant and whether among all of the vegetation in Sicily he might "not discover the archetypal one? There must certainly be one" (17 April 1787) [ob ich nicht unter dieser Schar die Urpflanze entdecken könnte. Eine solche muß es denn doch geben] (FA, I, 15, pt. I: 285–86). This interest in botany that developed in Italy increased upon his return and resulted in his lifelong research in the field.

Goethe's interest in the archetypal plant is also significant because it marks for many scholars the beginning of German classicism. It was a discussion about this plant that finally brought Goethe and Schiller together and solidified their friendship. Although Schiller had been living in the same small town as Goethe, Goethe had generally avoided contact with him. Goethe objected to the exaggerated sense of sentimentality in Schiller's early works and objected to Schiller's Kantian emphasis on the subject in his more philosophical essays. Goethe even refers to their viewpoints as representing two opposing poles of the earth. One day in 1794, both men attended a lecture and afterward struck up a conversation about the metamorphosis of plants. They then fiercely argued about the existence of the archetypal plant. Schiller claimed, much to Goethe's annoyance, that the archetypal plant was just an idea, i.e., that it did not really exist, and Goethe maintained the existence of his morphotype. Goethe reports, however, how eventually these two opposing poles were reconciled enough to become good friends (FA, I, 24: 434–38).

Goethe, of course, never discovered a particular plant to represent all plants, and even his essay on plant metamorphosis indicates a shift away from focusing upon a static form and toward a more malleable one. His scientific essay argues not for the existence of one archetypal form for the plant, but for an archetypal organ, the leaf: he argues that all organs of the plant are really a modified form of the leaf.[6] The malleability of the leaf, in one sense, becomes symbolic of Goethe's scientific classicism and anticlassicism. Although he envisions an archetype that is similar to a classic Platonic form in that it is to be found in all plants and their organs, the essence of that form is not static. Instead, the very form is defined by its ability to change and transform through the process of intensification. The most important aspect of the leaf within the treatise is not its stable form, but its very malleability.

The main part of this essay examines the transformation of this organ through intensification into a myriad of different forms. During the essay, the organ of the leaf becomes more and more defined, articulated, and specialized. Through six alternating polar stages of expansion (Ausdehnung) and contraction (Zusammenziehung), the leaf gradually emerges from the seed in the form of paired cotyledons, eventually forms the calyx, and then forms the petals and the highly specialized male and female sex organs of the plant, whose union results in the production of fruit and seed. The leaf becomes an example of the fluidity of nature in general as well as of one organ in particular. Intensification is able to transform a simple form into very specialized and often quite beautiful ones. The whole process, in addition, is a microcosm of the created world within Goethe's creation myth: it combines the spiritual with the material, so that the process of creation requires the cooperation of both.

Goethe is very meticulous in his descriptions of the process of metamorphosis, in part because he wishes to convince the scientific community that he is correct about the transformation of the leaf. He also wishes to teach his readers about a philosophical (or some might say mystical) process. He attempts to use his accurate and verifiable natural observations as a springboard for a philosophical program. Schrimpf has argued that Goethe's conception of objective knowledge makes him the last didactic writer of the nineteenth century (141). Where the Romantics distanced themselves from the object and focused upon their internal subjectivity, Goethe still believed that human beings could learn from objects (140–41). His scientific works are not merely a vehicle for reporting his findings, but are an attempt to educate his readers about the way in which they should look at the world and their role within it: he attempts to make his findings inseparable from his philosophy. In the case of his "Metamorphosis," this means that as the leaf progresses and becomes more specialized and its sap becomes more refined, the language of the essay becomes more esoteric and mystical.[7]

Throughout the "Metamorphosis," the process of Steigerung is described as a step-by-step progression (stufenweise [nos. 6, 28], fortschreitende [no. 6], Stufenfolge [no. 10]).[8] The leaf undergoes several stages (Stufen, literally "steps" [nos. 38, 48, 73]) and takes several steps (Schritten [nos. 3, 73]) as it ascends its spiritual ladder (geistiger Leiter [no. 6]) and strives to reach its goal (Ziel [no. 38]). In his account of the leaf's development, Goethe focuses upon nature's inner drive to become more specialized and articulated as it seeks to transcend its material beginnings. He also calls our attention to the Protean qualities of nature: a form need not be static, but may change and transform at will. When the first paired leaves (cotyledons) emerge from the seed, they are mis-shapen (unförmlich) and filled with crude material (rohe Materie [no. 12]). In their first and unintensified state, one can scarcely recognize them as leaves, and they can be "scarcely distinguishable from the substance of the whole" (von der Masse des Ganzen kaum zu unterscheiden [no. 12]). Goethe emphasizes the materiality and the crudeness of this first form. This first form also reflects the heavy materials of earth and water to which it owes (schuldig) its first existence. It is only able to intensify its form once it is exposed to things less material and more spiritual: air and light (nos. 13, 24).

That Goethe links the plant's development with its environment is quite significant. Steigerung enables the plant to transcend the matter that first gave it the impetus to grow. After it begins to grow, the more material elements impede its progress, whereas air and light promote it. In other words, the process of Steigerung becomes reflected in the materials that foster it. Goethe's account of the growth of the plant harkens back to the four Greek elements of earth, water, air, and fire. His plant comes to represent a microcosm of the whole universe: he incorporates the four elements within the life of the plant, and one witnesses how the organic and inorganic natural realms work together. An organism may only grow and develop to the extent that it uses all of the elements of the inorganic world. Polarities, too, play a central role within the intensification of matter: the two more spiritual elements of air and light (fire) are the polar opposites of the two more material elements of earth and water.

In "Metamorphosis," Goethe further marks the connection between the material and spiritual aspects of plant development. As the plant's organ progressively develops, the leaves become more specialized and form into particular plant organs: calyx, stamens, pistils, etc. Their exterior intensification is matched by an interior one. The plant's inner juices correspond to the leaf's greater articulation. At each stage, the sap of the plant becomes more pure and fine (reiner, feiner [nos. 26, 28]) to match the growing refinement of leaf (e.g., nos. 27, 28). The material aspect of the plant therefore intensifies to match the intensified form of the plant.

As Goethe continues the account of nature's drive toward greater development, he emphasizes the process of intensification from visible to olfactory perspectives. The calyx further filters and purifies the sap so that the petals form from a high degree of refined expansion (no. 41). Even the color of the plant indicates an intensification as the leaf changes from green into another color (nos. 41, 44): "The beautiful appearance of colors leads us to the notion that the material filling the petals has attained a high degree of purity, but not yet the highest degree (which would appear white and colorless)" (12: 83) [und die schönen Erscheinungen der Farben führen uns auf den Gedanken daß die Materie womit die Blätter ausgefüllt sind, zwar in einem hohen Grad von Reinheit, aber noch nicht auf dem höchsten stehe, auf welchem sie uns weiß und ungefärbt erscheint] (no. 45).[9] The striving of the plant, according to Goethe, is outwardly the most visible during this transition stage, where some petals of the flower are still partially green. As the calyx (that part of a flowering plant that supports the petals) transforms into the corolla (or petals) of the plant, one often sees "at the tip, edge, back, or even on the inner surface of a part where the outer surface remains green" (12: 82) [an den Spitzen, den Rändern, dem Rücken, oder gar an seiner inwendigen Seite, indessen die äußere noch grün bleibt] (FA, 1, 24: 122, no. 40).

Goethe found this transitional stage to be so remarkable that he commissioned watercolors to illustrate this phenomenon. This stage not only supported his claim that the organ of the leaf metamorphosed from one organ into other, but it also was indicative of a plant's inner striving away from materiality and toward greater complexity. It therefore served as one important example for the striving not only of whole organisms, but also of individual parts within one organism. The mixed colors of a petal always (jederzeit) indicated a refinement (Verfeinung [no. 45])—a growing specialization in both form and inner matter. For example, in a tulip, one can often see how the green within the petal actually belongs to the stem (zum Stengel gehörig) and remains attached to it, while the red part is more free to raise itself up (no. 44). Twenty years later, in his *Theory of Colors,* he even more affirmatively describes this process as one of inner striving: "There are white flowers in which the petals have worked themselves through to the highest degree of purity, and colored ones in which this beautiful elementary phenomenon may be seen here and there. There are also those which have only partially worked themselves free from green to reach a higher level [step or stage]" (*12: 258) [Es gibt weiße Blumen, deren Blätter sich zur größten Reinheit durchgearbeitet haben; aber auch farbige, in denen die schöne Elementarerscheinung hin und wieder spielt. Es gibt deren, die sich nur teilweise vom Grünen auf eine höhere Stufe losgearbeitet haben] (FA, 1, 23, pt. 1: 206, no. 623). The language in this passage further emphasizes the plant's own striving. The plant is the active agent in its attempts to reach a higher stage of development. Or, as Cornell has characterized it, some plant organs "'anticipate' those higher up on the plant's ascent" ("Goethe on Plants," 41). Not all petals are as successful as others, as some still display the lower stage of development through the mixture of green (the color of the lower form of the plant) to its colored petals. Such examples are important to Goethe because they demonstrate the types of internal causes that are of no concern to mechanists: the striving of nature and its parts that leads to more complex—and less material—forms.

Faust thus is not unique in his striving, but Goethe attributes to plants and plant parts a striving that accounts for organic metamorphosis. These entities feel driven from within to transcend their former existence. Not all beings, however, progress. In some cases, such as the partially green petal, the individual is only partially successful in overcoming its previous state. In more severe cases, the plant does not intensify beyond its leafy state or, conversely, intensifies so quickly that it does not complete its cycle. Too much nourishment (too much of the more base material of water) hinders flowers from ever forming, while deprivation of nourishment brings about the development more quickly as the "uncontaminated" sap becomes purer (reiner) and stronger (kräftiger) (no. 30). The plant that is overwatered has lost, so to speak, the Faustian wager. It is so satisfied in the moment that it does not push forward toward a higher existence. Its being is that of simple existence. Conversely, the underwatered plant will never reach the highest developmental state, where its two most specialized organs meet.

The refinement of the plant that is not undernourished or overnourished, however, continues and culminates in the formation of two distinct and opposing sexual organs. The plant organ has at this stage become so specialized that it literally splits itself in two. Once the plant has reached this stage of heightened development, true anastomosis, or the physical growing together of parts or vessels, is no longer possible (no. 61). Like the Aristophanic myth, where the halves of the originally united circle creatures are separated and may reunite only during sexual intercourse, these halves are excluded from ever truly reuniting. In the Platonic dialogue, sexual intercourse is not primarily about procreation, but is about a longing for a former wholeness. Similarly, in Goethe's account, the main significance of the sexual union of the masculine and feminine parts of the plant is not reproduction. Instead, he emphasizes the process of Steigerung that culminates in the union of two intensified polarities. The inner material of these two highly organized organs has become so "pure" and "refined" that the sexual union between them is not simply a physical one, but a spiritual one as well: "By changing one form into another, it [the leaf] ascends—as on a *spiritual ladder*—to the pinnacle of nature: propagation through two genders [the two sexes]" (*12: 76, emphasis added) [und durch Umwandlung einer Gestalt in die andere, gleichsam auf einer geistigen Leiter, zu jenem Gipfel der Natur, der Fortpflanzung durch zwei Geschlechter] (no. 6).[10]

Several commentators have noted the philosophical significance of this moment. Gray believes that it is "probable" that Goethe "had in mind here something like the spiritual marriage of Boehme's system" (85). Cornell argues that this pinnacle is teleological and is one which is akin to Goethe's *Faust* in that the organism is "over-reaching its lower existence" ("Faustian," 484). Kirby suggests that in this moment "the opposition of succession and simultaneity" are overcome, not only in thought, but also "through the activity of nature" (74). Portmann writes of this moment of union, which occurs at the plant's most concentrated phase, as the "closest one can come to the transcendence of space and time, a spiritual event, even a symbol of the union of God and man" (142). Krell sees a link between Goethe's and Kant's characterization of sexual union as a form of communication, but notes that what for Goethe is necessarily spiritual is treated as a contagion by Kant (4–5).

Although the result of the union of the sexes is regeneration, the essay shows that regeneration in itself is not the miracle of this union. Goethe gives numerous examples of the plant's ability to procreate apart from sexual union, such as when it reproduces asexually through eyes, stem cuttings, or gemmae at any stage of their development. He emphasizes the reunion of the masculine and the feminine, not then just because of the plant's ability to reproduce, but because the sexual union represents a reunion of intensified polar halves, alongside a reunion of their more symbolic characteristics (material and spiritual, space and time, etc.).

This reunion, moreover, is the very kind of reunion that he describes in his "Polarity" essay.[11] He characterizes the moment as a kind of creative act, one which is both physical and spiritual. The "third" thing that emerges from this union is both a seed (in the lower sense) and a special moment in time (in the higher sense): "Thus we are inclined to say that the union of the two genders is anastomosis on a spiritual level; we do so in the belief that, at least for a moment, this brings the concepts of growth and reproduction closer together" (12: 86) [so sind wir nicht abgeneigt, die Verbindung der beiden Geschlechter eine geistige Anastomose zu nennen, und glauben wenigstens einen Augenblick die Begriffe vom Wachstum und Zeugung, einanader näher gerückt zu haben] (no. 63). The moment is spiritual because physically the masculine and the feminine parts do not actually touch.[12] The masculine part of the plant emits pollen that then joins with the feminine part. Even this pollen, however, has lost much of its materiality. Pollen in its dustlike state is a very rarefied and refined version of the plant's baser, heavier juices. It has intensified alongside the development of the form of the leaf.

A close look at the language and tone of "Metamorphosis," especially in these passages involving procreation, reveals that, for Goethe, this essay is much more than a treatise on the development of a plant. It is also a philosophical treatise that addresses several forms of creativity, including procreation, organic productivity, and even scientific productivity. It therefore presents one of the most succinct versions of his natural philosophy. His use of nonscientific language in this essay, however, has alienated many scientific readers who believe that it merely contains poetical musings. Goethe, however, is using this essay to show the connection between his philosophy and observations of physical phenomena. His treatise bears many affinities with Plato's *Symposium* due to the way it portrays the relationship between the physical and spiritual. As already noted, his account of the sexual union of plant parts resembles Aristophanes' myth. In addition, Goethe's concept of Steigerung further mirrors the tradition established by Diotima's speech in that it sees a dual function to procreation and uses the image of a ladder to explain how one transcends the material realm.[13] These similarities become more evident through a comparison with Diotima's speech. According to her speech, one can procreate in one of two opposing ways: physically or spiritually. Those who procreate physically raise families "in the blessed hope that . . . they will keep their memory green through time and eternity." Those who procreate spiritually "conceive and beget" wisdom and related "virtues": "it is the office of every poet to beget them, and of every artist whom we may call creative" (208e). She then describes how one needs to transcend the relationships of physical bodies. In describing the process of attaining wisdom, and specifically of absolute beauty, she uses the image of a ladder:

Starting from individual beauties, the quest for the universal beauty must find him [the student] ever mounting the heavenly ladder, stepping from rung to rung— that is, from one [body] to two, and from two to every lovely body, from bodily beauty to the beauty of institutions, from institutions to learning, and from learning in general to the special lore that pertains to nothing but the beautiful itself— until at last he comes to know what beauty is. (211c)

The quest for spiritual knowledge begins with the body and builds up from the physical to the spiritual. It begins with something very simple, one physical body. As the student climbs the rungs, he or she learns more and more complex and abstract things until reaching the final step of absolute beauty.

As Diotima's ladder begins with the physical in order to attain the spiritual, Goethe's ladder of plant development begins with the most material aspect only to rise above it to spirituality. Unlike Diotima's ladder, which portrays a simple step-by-step progression, Goethe's ladder is coupled with polarity. Each step of his ladder is part of an alternating polar process of expansion and contraction. In addition, the pinnacle of Goethe's ladder is the reunification of two heightened, polar opposites—the masculine and feminine parts of the plant.

While true physical growing together of the masculine and feminine parts is no longer possible once they have developed into two disparate organs, they do, in some sense, reunite. The fruit that develops is a product of the union of masculine and feminine. The moment that symbolizes the reunion also represents a kind of created product. By returning to Goethe's concepts of polarity and Steigerung, we can see why this union was so significant for him. This spiritual growing together, or anastomosis, is important because it symbolizes the union of polar opposites of spirit and matter and the masculine and the feminine, if only for the moment. The reunion of the masculine and the feminine represents the heightened creation that is an androgynous whole and is more significant than the androgynous whole of the seed, which is only a physical manifestation. This heightened reunion also bears two kinds of fruit: the physical one of progeny, and a spiritual one, which represents a wholeness, a completion of former division.

While such examples may tend toward the abstract, Goethe also includes much more concrete examples that link botanical processes with methodological and artistic ones. Throughout the essay, he uses the process of intensification as a model for scientific writing and his own philosophical progression. The process of the plant's development, therefore, is not only a metaphor for creation generally, but also for human artistry and science.[14] Most importantly, however, Goethe fashions his essay to mirror the stages of development of the regular, progressive plant that he describes. In addition, as is discussed in more detail in chapter 4, he also wrote a didactic poem, "The Metamorphosis of Plants" ("Die Metamorphose der Pflanzen"), which also compares the progression of plants with the development of an ideal human relationship.[15]

Steigerung and Scientific Writing

Throughout the "Metamorphosis of Plants," Goethe's poetic and philosophical language subtly draws parallels between a plant's intensification and a human being's attempts to flourish or to be productive. He does so, on the one hand, because he believes that the scientist's own created product, in this case an essay on plant metamorphosis, should reflect as closely as possible the actions and processes of the object being studied. The very structure of the essay, therefore, becomes a didactic tool to explain the process as a whole. On the other hand, he also believes that human beings are so intimately connected to nature and its processes that our creative endeavors are analogous to nature's. If nature needs to undergo a process of intensification in order to flourish, then human beings, whether scientists or artists, must do so as well. He makes this connection between the intensification of plants and human beings more explicit when he compares the process of creating a scientific essay to plant development. For example, he discusses the "Blätter" (leaves/ pages [no. 9]) of his own work and presents his essay in "Schritten" (steps [nos. 9, 84, 112]) that echo the step-by-step growth of the plant. He even states that the proof of "formed and developed botanical eyes" (ausgebildete und entwickelte Augen) rests in our own eyes (Augenschein [no. 100]).[16] Moreover, when introducing his essay, he tells us that retrogressive (rückschreitende) plants uncover (enthüllen, literally "dehusk" or "dehull") the secrets of progressive ones. Similarly, the structure of the essay mirrors the path of a retrogressive plant. After the essay has followed the regular steps of the progressive plant, it turns back (Rückblick) to examine the eyes of the plant, "which themselves lie hidden" [welche unter jedem Blatt verborgen liegen] (no. 84). And just as the essay presents one complete plant cycle, from seed to seed, so too does the structure of the essay present a cycle: it concludes by returning to the beginning and repeating the main points of the whole.[17] By going back to rephrase his main points, Goethe imitates the method of observation that he advocates: studying the forward and backward (the progressive and retrogressive) movements of the leaf (no. 120).

For Goethe, it was important that the scientist imitate nature's striving toward greater articulation as closely as possible. Steigerung is a difficult concept to explain, but one that Goethe believes is quite visible—once we are aware of the patterns—during the growth process of the plant. By imitating the process within the body and structure of the scientific text, he similarly desires to make us see and experience the process of intensification. The process is not conveyed merely by words, but by the structure of the whole. He does so because he believes that imitating nature's processes reduces the risk of oversimplifying them. The scientific process becomes, as a result, highly creative. At the end of the essay, he explicitly invokes the model of the plant both to advance his own scientific work and to criticize the work of others. In acknowledging his indebtedness to his predecessors (Ferber and Linnaeus), he hopes that, even if he has not yet cleaned/purified (gereinigt [no. 107; cf. nos. 28, 30]) all obstacles, his own essay will not remain fruitless (fruchtlos [no. 107]). Whereas a plant's final act is to create fruit and seed, Goethe hopes the step he has taken in writing his treatise will lead to the truth [und es wird sich bald entscheiden, ob der Schritt, den wir gegenwärtig getan, sich der Wahrheit nähere] (no. 112). He also believes to be able to progress (fortschreiten [no. 108; cf. no. 19]) in knowledge, where Linnaeus stayed in place (stehen blieb [no. 108, cf. no. 111]). He contends that a thorough comparison with Linnaeus would only hold him back (zu lange aufhalten). And just as a plant proceeds step by step to reach a goal

(Ziel [no. 38]), so Goethe sees that Linnaeus was hindered (hinderte) from stepping (schreiten) toward his goal (Ziel [no. 108]) because of a superfluity of data (nos. 109,111): Linnaeus looked at trees and drew inferences about plants. Moreover, his study of trees required years to see a complete cycle, whereas Goethe was able to study a plant's cycle in one season. In other words, Linnaeus, overladen with material data, is like an overwatered plant that cannot progress. Goethe, however, is able to progress beyond Linnaeus's work and be productive to the extent that he follows the same dynamic, intensified pattern as a progressive plant.[18] Through a proper balancing of theory and data, he claims to have purified (gereiniget) these previous obstacles so that his essay is able to bear the scientific "fruit" of truth (nos. 107, 112). The plant, in essence, has provided the model for scientific research: too much water, as too much data, prevents the individual from progressing. Scientific theories advance, in this case, to the degree that they are modeled upon the progressive plant. Goethe even promises that, in the future, his scientific method will continue to follow the plant's path of intensification. He pledges to lay out his materials in a successive order (to match the successive development of a progressive plant) to make his theory more evident.

In a later essay (1820), Goethe overtly acknowledges that he fashioned his essay according to nature's model:

Moreover, in describing plant metamorphosis I found it necessary to develop a method which conformed to nature. There was not latitude for error as the vegetation revealed its processes to me step by step. Without interfering, I had to recognize the ways and means the plant used as it gradually rose from a state of complete encapsulation to one of perfection. ("The Influence of Modern Philosophy," 12: 28)

[Fernerhin bei Darstellung des Versuchs der Pflanzen-Metamorphose mußte sich eine naturgemäße Methode entwickeln; denn als die Vegetation mir Schritt für Schritt ihr Verfahren vorbildete, konnte ich nicht irren, sondern mußte, indem ich sie gewähren ließ, die Wege und Mittel anerkennen wie sie den eingehülltesten Zustand zur Vollendung nach und nach zu befördern weiß.] ("Einwirkung," FA I, 24: 442)

Goethe models his essay on nature because he believes that scientists who follow nature's model are less likely to be led astray in their work. They do not seek to "interfere," i.e., create their own theoretical models, but they strive instead to model their work upon the objects that they are studying. He attempts to let the object of study speak for itself. He then tries to replicate what he has seen, not by symbols or by mathematical formulas, but by imitating the very dynamic pattern within the structure and content of the whole essay. Once he has grasped the object through his "objectively active" thinking process, he brings to the reader's "spiritual eye" the meaning of plant metamorphosis. In this, the earliest of Goethe's published scientific works, he has already established one of the main points of his natural philosophy: human beings should imitate nature's processes as closely as possible in their study of them.

Goethe's account of the metamorphosis also demonstrates how he used the principle of polar reconciliation within his theory of metamorphosis. He states several times in his notes that he opposes epigenesis and preformation, the two main theories of development that were predominant during his time.[19] Although most late-eighteenth- and early-nineteenth-century discussions of sexual union and embryology took sides between the theories of epigenesis and preformation, Goethe found both principles too limiting, and he preferred to incorporate aspects of both theories into his scientific works ("Botanik als Wissenschaft" [FA 1, 24: 94–95]; "Gesetze der Pflanzenbildung" [FA 1, 24: 99]). Within "Metamorphosis," Goethe's emphasis on the form of the leaf would place him on the side of the preformationists, while his emphasis on the metamorphosis of form would place him among the epigenesists.

In a later essay, Goethe carefully distinguished himself from the prominent epigenesist, Caspar Friedrich Wolff. He points to three differences between his theory of metamorphosis and Wolff's. First, he thought that Wolff's attention to microscopic studies prevented him from seeing the analogies between plants and animals. Second, unlike Wolff, he believed that polarity explained how plant development occurred. And, finally, where Wolff thought the plant undergoes a process of degeneration as it readies itself for sexual reproduction, Goethe saw it as an intensified process toward perfection. Wolff was so invested in his physical explanation for organic formation that he believed plant forms degenerated as they shed their materiality ("Wenige Bemerkungen," 1817 [FA 1, 24: 432–33]). Each of these differences points to important aspects of Goethe's natural philosophy. The first two (that all of nature is related and that polarity is one of nature's main forces) have already been discussed. The final one, which links a plant's intensification to a process of perfection, is a topic in this chapter and is discussed again in chapter 4.

Goethe interweaves his botanical discoveries with his natural philosophy in the form of a didactic essay. This integration, moreover, is at the core of his philosophy.[20] Many of his scientific works have the same didactic model. He returns again and again to the idea that human endeavors, scientific or philosophical, are successful if human beings imitate nature's creative model. According to Goethe, this model does not deny the importance of the material side of the universe. It incorporates the material within the path toward development. The plant never can totally abandon its need for water and earth. But too little water will force a plant to blossom prematurely and render it sterile. Too much water will prevent the plant from ever finding its greatest articulation of form and will prevent it from bearing fruit. The flourishing plant must have the proper amount of water to sustain the kind of developmental growth fostered by the more ethereal air and light.

So far, this chapter has examined the process of intensification in the animate realm. Goethe, however, also sees evidence of striving in the inanimate realm. According to him, certain colors intensify from seemingly inner processes. In addition, just as the image of the ladder serves as a metaphor for the principle of Steigerung in "The Metamorphosis of Plants," so too does Goethe turn to the image of the ladder in The Theory of Colors to explain the progressions of color, scientific methodology, and intellectual growth.

The Theory of Colors
The Scientific Methodology of Steigerung

Twenty years after writing the "The Metamorphosis of Plants," when Goethe was working upon his Theory of Colors (1810), he once again patterned his scientific writing upon the object of his study. Within The Theory of Colors, he describes how pure red arises from a ladderlike series of progression. (According to Goethe, pure red forms after yellow darkens to orange-red, which then meets a blue that has intensified to violet.[21]) Similarly, the structure of the Didactic Section of The Theory of Colors is based upon an image of a ladder that begins with the physical (the organ of the eye) and gradually ascends toward the spiritual (the moral and symbolic uses of color). The "ladder" of this section, just like the ladder of "Metamorphosis," contains six progressive steps.[22] He structures his argument according to a ladder for two reasons: first, he believes that the proper scientific methodology requires that we start with what is simple and then ascend toward what is more complex; and, second, he thinks that scientists should model their work upon the objects that they study. The structure of this text once again mirrors its contents: the text is modeled according to the very principles that it is discussing.

Where Schöne has argued that one should read *The Theory of Colors* as a literary text because of its rich symbolism (9), it might be more accurate to say that Goethe, by closely combining the meaning of the text with its structure, meant to create a philosophical text based upon nature's processes.[23] Goethe intends his treatise, therefore, to educate his readers not only about color, but also about the scientific method as a whole. The whole work is structured to encourage his readers to do the experiments to learn firsthand about the principles underlying them. Just as Aristotle's *De Anima* teaches about the activity of the mind (*nous*) by forcing the reader to use it (esp. 429b10–25, 430a10–30, 431a1–5), so Goethe hopes that the structure of *The Theory of Colors* will aid his readers in understanding its content.

The titles of the six parts of the Didactic section already indicate the progressive nature of the whole:[24] Physiological Colors, Physical Colors, Chemical Colors, General Inward Views (Allgemeine Ansichten nach innen), Neighboring Relationships (Nachbarliche Verhältnisse), and the Sensory-Moral Effect of Color (Sinnlich-sittliche Wirkung der Farbe). The very ordering of these sections is once again similar to Diotima's ladder, which begins with a solitary, physical body and concludes in the philosophical realm.

A brief overview of these parts illustrates the structural function of Steigerung. The first part is subjective and deals with the eye of an individual. The second part expands the scope of the investigation beyond the simple perception of color by the eye and treats color as a "transient effect on light" (no. 688). In this section, Goethe examines colors that arise as light passes through various media, whether through varnishes, prisms, the atmosphere, etc. The third part, focusing upon chemical colors, further expands the discussion to include both the organic and inorganic realms. The role of polarities is revisited from the first section, but this time they act together with Steigerung. He stresses nature's unity, by correlating intensified colors with mature plants (nos. 619–35) and highly developed animals (nos. 647, 657).

The fourth part continues the expansion of the previous sections by connecting phenomena that he had treated separately before. Whereas the first part of the text openly treated polarities and Steigerung, Goethe now informs his readers that he also has organized the structure of the text around these two main principles. Recapitulating the first three parts, his language recalls "Polarity" and "Reunion" by emphasizing separation and union:

Until now we have almost forcibly kept separate phenomena which constantly sought to satisfy their nature and the needs of our intellect by reuniting.

In this continual sequence we have sought to define, separate, and organize the phenomena as much as possible. No longer in fear of mixing or confusing these phenomena, we may now show the general principles we can deduce for them in the closed circle of our subject, and then indicate how this circle is connected to other phenomena in nature, how they are linked together. (12: 266)

[Wir haben bisher die Phänomene fast gewaltsam aus einander gehalten, die sich teils ihrer Natur nach, teils dem Bedürfnis unsres Geistes gemäß, immer wieder zu vereinigen strebten.] (no. 688)

[In dieser stetigen Reihe haben wir, so viel es möglich sein wollte, die Erscheinungen zu bestimmen, zu sondern, und zu ordnen gesucht. Jetzt, da wir nicht mehr fürchten, sie zu vermischen, oder zu verwirren, können wir unternehmen, erstlich das Allgemeine, was sich von diesen Erscheinungen innherhalb des geschlossenen Kreises prädizieren läßt, anzugeben, zweitens, anzudeuten, wie sich dieser besondre Kreis an die übrigen Glieder verwandter Naturerscheinungen anschließt und sich mit ihnen verkettet.] (no. 689)

Both nature and the human mind have the same impulses: both wish to reunite separated wholes. In the natural realm, Goethe is referring to such things as the complementarity of color. In the intellectual realm, he is referring to two disparate scientific approaches. He believes that once one has progressed to this fourth level, one can act like the creator in "Reunion" and reunite what was once separated. He wishes to reunite two once separated scientific approaches, thereby creating his own scientific whole. His discussion begins with separations, but then, because the very nature of mind (as of nature as a whole) craves wholeness, it brings about syntheses. His theoretical method of practicing science reflects the workings of the eye. Once the eye is presented with one phenomenon, it craves and then creates the opposite one. Burwick similarly points out that Goethe divides colors into three categories, where the third acts as a combination of the prior two: "In Goethe's division, the physiological colors are subjective, the chemical colors are objective, and the physical colors derive from subjective-objective interaction" (68).

The fifth section significantly expands the discussion by arguing for the relevance of the theory for different fields. Goethe begins by advising philosophers to participate in the principle of Steigerung. In this passage, he instructs scientists first to distinguish themselves from the world and from the objects of their study. Then, however, they must be able to come back together and reunite with the world in a higher sense [und mit ihr [der Welt] wieder im höhern Sinne zusammenzutreten] (no. 716). He instructs those who study nature to view the world objectively and subjectively to have a comprehensive understanding of it. One cannot remain fixed in one perspective. One must be as malleable as the eye.

In a succinct version of his criticism of modern science, Goethe concludes this section by warning scientists to be careful, when using symbols or mathematical formulas, not to replace or confuse "the thing with its sign" (no. 754).[25] Although metaphors or formulas act as important aids to the scientists in their inquiries, problems arise when they confuse the thing with its symbolic representation. One problem with turning to a formula to describe nature is that one soon forgets its other, nonmathematical aspects and treats the world as if it acts only according to formulas. In the case of Newton, Goethe argues that Newton was so interested in the angle of refraction, he no longer discussed the actual phenomenon of color. Goethe further complains that mechanical formulas tend to "transform living things into dead ones" (no. 752): scientists treat natural entities as if they were machines. To avoid the problems of confusing the sign with its object and "killing" the object, Goethe advocates beginning with the simplest phenomena (such as light and darkness) and treating the phenomena themselves as a dynamic formula (such as polarity). From there, researchers are to develop more complex formulas based upon the simpler ones (no. 755), much in the way in which Goethe begins the book by discussing polarities before progressing to the concept of intensification. In other words, as Goethe on the large scale tries to imitate nature's dynamic characteristics within the structure and content of his own scientific texts, on the small scale he wants his readers to replicate as closely as possible nature's activities. Indeed, Goethe informs us that the best scientific sign is one "in which the basic sign expresses the phenomenon itself" (no. 756). Such a sign, according to Goethe, is polarity (nos. 756–57), which avoids the problems of mathematical formulas because we are never forced to separate the concept from its functioning.

The scientific "gains" of Goethe's principles are not the practical ones of a more mechanistic approach. (Indeed, he does not often even admit to the great power and usefulness of mathematical and mechanical approaches and thereby makes his own science seem quite partisan and one-sided.) One of the goals of his science is to bring about an understanding of nature that has philosophical consequences for scientists. This goal is most clear in the last section of the Didactic part. Like the pinnacle of Diotima's ladder, Goethe emphasizes the aesthetic aspects of this final rung ("Anzeige," FA, 1, 23, pt. 1: 1047). Here he discusses the sensory-moral effects of colors and includes discussions of symbolic, mystical, and religious uses for color. He asserts that scientists may gain a sense of harmony and well-being by conducting the experiments, so that they not only study a whole, but also seek to achieve a philosophical wholeness within themselves.[26]

Goethe claims that scientists, as the observers of colors, are the recipients of the active influence of colors, i.e., they are passive in that they are being acted upon. This passive state, as discussed in chapter 1, is a limitation (Beschränkung), from which one is liberated through activity. In reacting and acting, the eye creates a satisfying whole (eine befriedigende Ganzheit [no. 812]). Similarly, the scientist, observing colors, begins the experiment in a passive position. The colors of the color wheel have a pleasing effect upon the observer (no. 815). Although all of the colors exist naturally, a circle that simultaneously exhibits all of the colors does not naturally exist. (Goethe specifically points out that a rainbow is incomplete because it does not contain the color pure red [no. 814].) By placing the various experiments side by side, however, the scientist is able to produce a perfectly beautiful effect that exists only through human creativity (no. 815). This created effect, this experiment, gives us an idea of harmony (Idee dieser Harmonie) that we can then feel in our soul/mind (im Geiste gegenwärtig fühlen [no. 815]). The experiment begins in the realm of the physical, but then transcends the physical of both the object (color) and the subject (body). Goethe argues that scientists, by bridging the gap between subject and object, create a feeling of harmony within themselves. Through artifice, the very scientific method, they create a whole that is pleasing to the eye and the mind. Goethe makes no excuse here, as elsewhere in this text, that his scientific claims go beyond the accepted boundaries of science. He simply suggests that, if we do the experiments, we will have the proof: we will be able to experience these responses for ourselves.

The entire work concludes with reflections upon the effects of specific colors on human beings. The closing paragraphs return to the question of the relationship of the sign to the symbol. Some allegorical associations of colors (as when we associate hope with green) are merely conventional. He maintains that others, such as the association of majesty with pure red, are not conventional at all, but are in complete harmony with nature. He believes that all who see this color automatically associate it with an elevated feeling. Here he cites as evidence that different peoples at different times have used a purple-red to symbolize majesty. For Goethe, the symbolic distinction between green and pure red is significant. These two colors illustrate that he attempted in his philosophy to include an understanding of relativism and natural law that coexist side by side. The meaning of green is not stable or universal, but is transient, as are all earthly things. The meaning of pure red for him is akin to a natural law that is true everywhere and at all times. It is able to do so because the sign is the same as the thing.

Goethe concludes the whole work with an overt appeal to mystical symbols. His discussion of the mystical use of color mirrors the creation myth in both his autobiography and "Reunion." He directly links the two triads—blue, yellow, and green; and blue, yellow, and pure red—with mystical and philosophical traditions. Blue and yellow create the material world (green) and the spiritual one (pure red):

We must grasp how yellow and blue diverge, and should reflect especially on the intensification in red where the opposites incline to one another and merge to create a third element. Then we will certainly arrive at the mystical and intuitive perception that a spiritual meaning can be found in these two separate and opposite entities. When we see them bring forth green below and red above, it will be hard to resist the thought that the green is connected with the earthly creation of the Elohim, and the red with their heavenly creation. (12: 296)

[Wenn man erst das Auseinandergehen des Gelben und Blauen wird recht gefaßt, besonders aber die Steigerung ins Rote genugsam betrachtet haben, wodurch das Entgegengesetzte sich gegen einander neigt, und sich in einem Dritten vereinigt; dann wird gewiß eine besondere geheimnisvolle Anschauung eintreten, daß man diesen beiden getrennten, einander entgegengesetzten Wesen eine geistige Bedeutung unterlegen könne, und man wird sich kaum enthalten, wenn man sie unterwärts das Grün, und oberwärts das Rot hervorbringen sieht, dort an die irdischen, hier an die himmlischen Ausgeburten der Elohim zu gedenken.] (no. 919)

Because the same elements (colors) create both opposites, one cannot say that one side (the passive blue or the active yellow) is more important than the other. Because the spiritual side has intensified, however, it is more developed than the material one. Again, pure red and green are opposites: one representing a natural and divine relationship, and the other a conventional and human one. Green and red are also complementary colors to each other, i.e., if one stares at a green object long enough, one will soon see a red afterimage and vice versa. These two opposites are then further reconciled in a mysticism which combines human perception (Anschauung) with nature (Natur).

By now turning to Goethe's discussions particularly of red and green, one can gain a clearer idea of why he turned to such mystical symbolism within a work that he hoped would start a new scientific revolution. Once again, Steigerung is the force behind progression.

The Red and the Green

According to Goethe's natural philosophy, the very same forces that were at work in the plant's intensification of form are present in the development of certain colors: inanimate nature strives to transcend its materiality much in the same way as organic nature. By comparing Goethe's treatment of the intensified polar reunion of pure red with the nonintensified one of green, one can further trace the importance with which he endowed intensified polar unions. He equates green with physical reproduction or that kind of reproduction related to the seed of a plant, and red with metaphysical production or the creativity that arises from a union of intensified opposites. Both creation and procreation are illustrative of nature's productive essence, but red represents nature's drive toward transcendence.

Red and green, because they combine the polar characteristics of the "mother colors" (Mutterfarben) yellow and blue, are symbolically more important than other colors in Goethe's theory. They both arise from the union of yellow and blue—green with the simple, physical mixing of the two, and red through the result of an intensified union after yellow darkens to orange-red and meets blue that has darkened into violet. In his introduction, Goethe explains the relationship of these two colors to the entire color wheel:

For the moment we will take only a brief look ahead with the statement that light and dark, brightness and darkness, or, to use a more general formulation, light and nonlight, are necessary for the production of color. The color we find emerging closest to light we term yellow; a second which arises closest to darkness we call blue. A perfectly balanced combination of these two colors in their purest form will produce a third color we will call green. Each of the first two colors may also create a new phenomenon of its own when concentrated or darkened. They take on a reddish appearance which may be intensified to such a degree that the original blue and yellow become almost impossible to recognize. (12: 165)

[Gegenwärtig sagen wir nur so viel voraus, daß zur Erzeugung der Farbe Licht und Finsternis, Helles und Dunkles, oder, wenn man sich einer allgemeineren Formel bedienen will, Licht und Nichtlicht gefordert werde. Zunächst am Licht entsteht uns eine Farbe, die wir Gelb nennen, eine andere zunächst an der Finsternis, die wir mit dem Worte Blau bezeichnen. Diese beiden, wenn wir sie in ihrem reinsten Zustand dergestalt vermischen, daß sie sich völlig das Gleichgewicht halten, bringen eine dritte hervor, welche wir Grün heißen. Jene beiden ersten Farben können aber auch jede an sich selbst eine neue Erscheinung hervorbringen, indem sie sich verdichten oder verdunkeln. Sie erhalten ein rötliches Ansehen, welches sich bis auf einen so hohen Grad steigern kann, daß man das ursprüngliche Blau und Gelb kaum darin mehr erkennen mag.] (FA, I, 23: 26–27)

This passage echoes the language of the "Polarity" essay. Goethe postulates two kinds of unions: a simple, unintensified one (green) and a more complex, intensified one (a type of red). The appearance and effect of red are the very definition of intensification because it is the union of two progressively intensified (fortschreitende Steigerung [no. 523]) halves. Red forms when blue and yellow meet in their most intensified states. Green is by definition a balanced, physical mixture of the two most simple colors, yellow and blue. It is only able to link (verknüpfen) the gulf (Kluft) between the polar opposites of yellow and blue. True mediation (die wahre Vermittlung) occurs only through pure red (purpur [no. 539]).[27]

The unions that result in the colors green and pure red are significant aside from the fact that they arise according to two different principles. Their unions also symbolize the balancing or the reconciling of two sets of opposed characteristics that go beyond the visible appearance of color. Goethe writes that the colors inherently represent a variety of qualities:

Plus.	Minus.
Gelb (yellow).	Blau (blue).
Wirkung (cause).	Beraubung (deprivation).
Licht (light).	Schatten (shadow).
Hell (brightness).	Dunkel (darkness).
Kraft (power).	Schwäche (weakness).
Wärme (warmth).	Kälte (coldness).
Nähe (nearness).	Ferne (distance).
Abstoßen (repulsion).	Anziehen (attraction).
Verwandschaft mit	Verwandtschaft mit
Säuren (affinities with acids).	Alkalien (affinities with alkalis).

(no. 696)

He speaks throughout his work of the symbolic meanings of colors and their activities. Their interaction, as the foregoing list shows, comes to symbolize not only the formation of color or even visible phenomena, but all phenomena. And while one may readily see the associations that he attributes to each side in this list, his applications of these characteristics to green and pure red are not that clear, but require explanation and—in some cases—an apparent leap of faith. And no matter how much he tries to argue that subjective and emotional responses are just as much a part of science as are quantifiable facts, his presentation of these colors remains cryptic. He merely sets forth the proposition that most colors lean more toward one side in this list. In his conception of colors, green and pure red are unique in that they take on the characteristics of *both* sides: red by reconciling them, and green by balancing or neutralizing one set against the other. Green and red, representing polar unions, therefore symbolize two opposite microcosms: green, the microcosm of the physical world, and red, the more inward, intensified one. And, as human beings react in different ways to different colors, they too become participants in the process of polar unions and intensification.

Goethe considers green a lower combination than red for several reasons. First, yellow and blue are at their lowest, or most simple, state when they combine to form green (no. 745). Second, and apparently directly resulting from the first, the mind's reaction to green is not as developed as its reaction to pure red. He argues that our philosophical and emotional reactions to the color are closely linked to its physical manifestation. In the most perfect green, when yellow and blue meet in perfect equilibrium, the eye and the mind (Gemüt) rest upon it as something simple (auf einem Einfachen). One's eye as well as one's mind reach a state of rest: "One cannot and will not go beyond it" (283) [Man will nicht weiter, und man kann nicht weiter] (no. 802). The perception of the color has a direct influence not only upon our mood, but upon our intellectual state of mind.

Goethe further links green with certain symbolic or allegorical associations. He associates green with earthly (irdisches) aspects that are associated with physical union (no. 919), i.e., physical begetting. Green demonstrates that yellow and blue together give birth (Erzeugung) to a unity that is completely apart from themselves. He states that such a relationship is satisfying in a very real way (reale Befriedigung [no. 802]) seemingly because it is tangible in the same way as a seed is a tangible symbol of the plant's reunion of parts. The seed is symbolic of the plant's attempt to participate in immortality: it will live on, in a manner of speaking, through the seed. Once one reaches this kind of real satisfaction as symbolized by green, it becomes difficult to do something else.[28] Even Mephistopheles remarks that while all theory is gray, "the golden tree of life is green" (Faust, 2038).

Green, then, is not that dissimilar to the biblical command to be fruitful and multiply. It represents, for Goethe, a type of immortality in that reproduction enables the continuation of the line, if not of the individual. Green does not incite one to go beyond it, but leaves one self-satisfied. It represents the satisfaction that Faust managed, through his striving, to avoid. If "Human beings err, as long as they strive" [Es irrt der Mensch, solang er strebt] (Faust 317), then while green might prevent people from falling into error, its complacency is also not what is the highest about us. Faust's saving grace, so to speak, is his desire to go beyond the present experience and never to rest in the moment. So, too, in *Elective Affinities*, Charlotte eventually acknowledges the harm in her adherence to family values. When Ottilie accidentally kills Charlotte's child, Charlotte realizes that the tragedy could have been prevented had she been more flexible and granted her husband a divorce. In the novel, Goethe does not so much criticize the family itself but the belief that it is more important than other aspects of our lives. Similarly, it is not a bad thing that an overwatered plant keeps growing without progressing. But it is also not the highest level of existence that a plant may achieve. The transition from green to red in some flower petals marks for Goethe a moment of striving toward the next level of existence ("Metamorphosis," nos. 40–44).

Goethe turns to pure red as a means of illustrating that there is much more to nature's striving than procreation. His discussion of pure red (purpur, literally "pure-pure" [nos. 612, 703]) concentrates upon its apparent elevatedness and purity. The process of intensification that leads to red, whether as a result of prismatic or chemical processes, is, according to Goethe, more tenuous (no. 531) than the physical mixture of green. Although green is "a beautiful and pleasant phenomenon" [ein schönes und angenehmes Phänomen], he maintains that red is "more graceful" [eine anmutigere Farbe] and is "the culmination of the process" [der höchste Punkt der ganzen Erscheinung] (no. 702). Like the spiritual anastomosis of the plant's two sexual organs, red represents a union of opposites at their most developed stages. In his comparisons of the two colors, he stresses what he believes to be the greater complexity of red. Green is a unity (Einheit), red a generality (Allgemeinheit). Whereas green points to a duality (Dualität), red points to harmony (Übereinstimmung [no. 698]). Moreover, in the closing paragraphs of the work, Goethe links pure red to heavenly creativity, and green to an earthly one (no. 919).[29]

Red, like green, represents a kind of satisfaction, but a higher, more ideal one (ideale Befriedigung [no. 794]). And while Goethe never establishes—beyond appealing to our emotional reaction to red—his justification for its philosophical elevatedness, he is more explicit about its symbolic function. The result of the union of opposites brings forth not children, but activities that may then bring about poetic or artistic creations. Goethe throughout his corpus often discusses various mani-festations of immortality that arise through a drive to be active. In a con-versation with Eckermann, he describes how the belief in his own im-mortality springs from the concept of activity [Tätigkeit], which he further characterizes as a kind of ceaseless, yet directed or effective, ac-tivity:

> To me, the eternal existence of my soul is proved from my idea of activity; if I work on incessantly till my death, nature is bound to give me another form of ex-istence when the present one can no longer sustain my spirit. (4 February 1829)

> [Die Überzeugung unserer Fortdauer entspringt mir aus dem Begriff der Tätigkeit; denn wenn ich bis an mein Ende rastlos wirke, so ist die Natur verpflichtet, mir eine andere Form des Daseins anzuweisen, wenn die jetzige meinem Geist nicht ferner auszuhalten vermag.] (FA, 2, 12: 301)

For human beings, pure red represents a creative activity that leads to a pregnancy of mind, rather than a pregnancy of body. For nature as a whole, pure red comes to symbolize an active striving beyond simple ex-istence, as when a plant intensifies its form of the leaf or when individual entities strive to create entirely new forms for themselves (chapter 3).

Red also exhibits a stronger inner drive than green. Whereas green requires a physical mixing from without to bring about its existence, red appears to be driven from within to intensify:

> It will come as no surprise that a genuine resolution occurs in the union of the in-tensified poles, a satisfaction in the ideal realm. Among the physical colors, then, this most exalted of color phenomena arises from the merger of two opposites which have gradually prepared themselves for union. (*12: 282)

> [. . . so läßt sich denken, daß nun in der Vereinigung der gesteigerten Pole eine eigentliche Beruhigung, die wir eine ideale Befriedigung nennen möchten, statt finden könne. Und so entsteht, bei physischen Phänomenen, diese höchste aller Farbenerscheinungen aus dem Zusammentreten zweier entgegengesetzten Enden, die sich zu einer Vereinigung nach und nach selbst vorbereitet haben.] (no. 794)

In their path toward intensified union in red, the colors are very active. Unlike green, which represents a kind of stasis, red, as well as the process toward it, is a striving heightening (strebende Steigerung [no. 794]) that represents motion and activity. This momentum toward condensation, saturation, and darkening comes from within the colors themselves (nos. 517, 794).

Intensification toward red seems analogous to Lucretius's swerve of the atom: an inner drive that can be found in the smallest inorganic units. Specifically, Goethe's emphasis on the self-motivation of colors mirrors Lucretius's claim that atoms spontaneously move. Just as Lucretius links inanimate and animate nature through free will (the spontaneous swerve of the atom), so too does Goethe link both realms through striving and polar reconciliation. Red represents a reconciliation of both poles, even if only one color has heightened, because yellow and blue, in their heightened state, are the same. Red, therefore, also once again illustrates that opposites in Goethe's natural philosophy may simultaneously exist without canceling each other. Moreover, red contains all other colors, "in part manifest, in part latent" (teils actu, teils potentia [no. 793]). The reconciliation of opposites further proves to be a heavenly (himmlisches) phenomenon as opposed to the earthly one of green. Pure red in this respect is akin to Goethe's most famous account of heavenly reconciliation in the ending to *Faust II*. Although the ultimate reconciliation of opposites—not only the two souls of Faust's breast but also his reunion with Gretchen (11,954–65)—does not occur while he is alive, his striving for such reconciliation during his lifetime guarantees it in the next.

Goethe is perhaps least convincing as a scientist when he turns to the vocabulary associated with spiritual, natural striving throughout his *Theory of Colors*. He, however, is aware of this problem and even partially addresses it. In the concluding paragraph of the work, he acknowledges that such mystical statements will "expose" him "to suspicions of wild imaginings" (Verdacht der Schwärmerei [no. 920]). He therefore appears to step back from his previous claims in order to insure a more favorable reception for his work. This step back, however, is but a momentary one. He ultimately hopes that a favorable reception "will enable allegorical, symbolic, and mystical applications and interpretations" (12: 296–97) [an allegorischen, symbolischen und mystischen Anwendungen und Deutungen . . . gewiß nicht fehlen wird] (no. 920). Ultimately, a favorable reception in his mind would be one that would mix the more clearly scientific aspects of the work with more symbolic ones.

Goethe, in part, does not explain his mystical, natural beliefs because he believes that they are inexplicable. One cannot express them because one can only experience or feel them (no. 918). He argues that these mystical meanings are just as much a part of nature as they are of intu-

itive perception (menschliche Anschauung [no. 918]): it is part of human nature to create symbols and therefore part of the investigator's
role to interpret them. Thus, a particular color or a particular symbol
may simultaneously represent numerous different meanings for different groups and still be part of "nature." A triangle, he notes, has a completely different meaning for a mathematician than it does for a mystic
and both meanings ought to be investigated (no. 918). He writes at the
very beginning of *The Theory of Colors* that one of his main aims was to familiarize the "friends of nature" with the "language of nature." And as
these "friends" may be scientists, artists, technicians, philosophers, etc.,
he attempts throughout the work to draw our attention to as many different meanings, including the aforementioned mystical ones. In the
process, however, as his closing paragraphs especially demonstrate, he so
expands the scope of his study that it no longer seems scientific at all.

The Poetry of Science

Both Goethe's *Theory of Colors* and "Metamorphosis of Plants" were
generally viewed by his public as poetic, rather than scientific, productions. The main reason was not, as Goethe himself believed, that people
were not aware that he had devoted a great part of his life to scientific
studies (FA, 1, 24: 752). A major stumbling block to Goethe's acceptance
as a scientist was the kind of science that he promoted. He himself, however, did not believe that he was either poeticizing or anthropomorphiz-
ing nature within his scientific works. Because he viewed nature as a dynamic, creative entity, he argued that one had to go beyond mathematical
formulas to describe it. He further claimed that his observations of nature led to a better understanding of human beings and their activities.
The aim of science was therefore not just to gain understanding about
nature, but also to gain an understanding of the role of human beings
within the context of nature.

Throughout his life, Goethe linked plants and plant development to
human beings and their endeavors.[30] For him, the study of plants, beginning with his early search for the Urpflanze, provides a source for understanding not only botany, but also all living things. In a letter to Charlotte von Stein (Rome, 8 June 1787), he describes the qualities of his
Urpflanze, "the most unusual creature in the world" (das wunderlichste
Geschöpf von der Welt), as a "model" (Modell) and a "key" (Schlüssel).

He writes in great excitement about how close he is to discovering the secret (Geheimnis) of plant reproduction and organization. This secret for Goethe goes well beyond "artistic or poetic shadows" (malerische oder dichterische Schatten), but represents an "inner truth and necessity" (innerliche Wahrheit und Notwendigkeit). He writes of how "the same law may be applied to all other living things" [Dasselbe Gesetz wird sich auf alles übrige Lebendige anwenden lassen].

Toward the end of his life, Goethe elaborated on this belief. Rather than limiting intensification to organic development, he also applies it to politics and aesthetics. In a conversation with Eckermann (13 February 1829), Goethe again speaks of great secrets (grosse Geheimnisse) that lie hidden in nature and promises to entrust to Eckermann some that he has already discovered. He also tells Eckermann that he will express himself somewhat strangely during the course of their conversation. A main part of these secrets is nature's drive toward refinement. He outlines how the growth process of a plant, as it grows from bud to bud and culminates in a flower, is the same process as the segmented bodies of caterpillars and the segmented vertebral skeletons of human beings culminating in the development of a head. Nor does this analogy between plants and human beings end in the physical realm. Goethe uses this principle to explain politics and artistry. At this point, the somewhat strange aspect of which he warns Eckermann comes into play. Goethe views communities as an organic whole. Like a plant, a community may intensify to bring about a very specialized member: a hero or a great artist. For example, he compares the culmination of the beehive in the queen bee to the culmination of the French poetic powers in Voltaire. This passage illustrates that the very basis for human artistry and greatness is the same natural principle that explains nature's striving. Human beings are a part of nature, and, as such, their flourishing and development, while on a more developed and complex scale, mirror that of plants. One may, therefore, also see how closely Goethe integrates human beings within nature as a whole.[31] Human striving for excellence in the arts has a parallel foundation in nature's striving for refinement.

Because Goethe so closely links our activities with those of nature, it becomes clear why he was so vehemently opposed to mechanistic approaches to science. If one reduces nature to mechanistic and formulaic principles, Goethe believes that it may become too easy to reduce human endeavors in the same way. He explains to Eckermann how his natural studies made certain aspects of human beings clear to him:

Without my attempts in natural science, I should never have learned to know mankind as it is. In nothing else can we so closely approach pure contemplation and thought, so closely observe the errors of the senses and of the understanding, the weak and strong points of character. (*Conversations with Eckermann*, 13 February 1829)

[Ohne meine Bemühungen in den Naturwissenschaften hätte ich jedoch die Menschen nie kennen gelernt, wie sie sind. In allen anderen Dingen kann man dem reinen Anschauen und Denken, den Irrtümern der Sinne wie des Verstandes, den Charakter-Schwächen und -Stärken nicht so nachkommen.] (FA, 2, 12: 308)

Goethe goes on to explain how individual human beings have to be able to raise up (erheben) their own abilities to reason, if they are ever to understand nature's processes. Again, he stresses that we must take nature as our model in this process. Otherwise, we will see only the dead and the static aspects of nature, not its malleability and flexibility.

Goethe took the model of a plant's progression beyond his scientific studies. The intensified plant served not only as a model for a scientific method or the progression of a particular individual, but also for several of his own works of art. At one time, Goethe considered basing his autobiography upon the model of plant growth. Early childhood would have been paired with the cotyledon stage, boyhood with lively green steplike growth, and youth with the rush of spikes and panicles to blossom (FA, 1, 14: 971).[32] He also compares his reason for abandoning the model to plant growth. Not all plants will flourish in a particular place or time. Perhaps in another place or time, this model for writing may have worked, but not now. He concludes the metaphor by once again comparing his work to fruit and explains that some of it ripens very slowly and some never ripens at all (FA, 1, 14: 972).

Goethe, however, returned to the image of an intensified growth process to explain the structure and the content of his "Novella." The novella is the story about an escaped lion that is eventual pacified and subdued through a child's flute playing. Eckermann, upon reading the novella, is slightly disconcerted. He finds the ending, which concludes in this poem, to be too ideal:

And so blessed angel bringeth
To good children help in need;
Fetters o'er the cruel flingeth,
Worthy act with wings doth speed.
So have tamed, and firmly iron'd
To a poor child's feeble knee,
Him, the forest's lordly tyrant,
Pious Thought and Melody.
(11: 280)

[Und so geht mit guten Kindern
Seliger Engel gern zu Rat,
Böses Wollen zu verhindern,
Zu befördern schöne Tat.
So beschwören, fest zu bannen
Liebem Sohn an's zarte Knie
Ihn des Waldes Hochtyrannen,
Frommer Sinn und Melodie.]
(FA, 1, 8: 555)

When Eckermann expresses his objections to the ending, Goethe argues that to have concluded the story otherwise would have been prosaic. Everyone can imagine the general return to normalcy after the lion has been tamed. He defends the concluding poem by arguing that the Steigerung of the whole required a more lyrical ending. He compares the climax of this work with that of a flower:

As a similitude for this novel . . . imagine a green plant shooting up from its root, thrusting forth strong green leaves from the sides of its sturdy stem, and at last terminating in a flower. The flower is unexpected and startling, but come it must—nay, the whole foliage has existed only for the sake of that flower, and would be worthless without it. (18 January 1827)

[Um für den Gang dieser Novelle ein Gleichnis zu haben, . . . so denken Sie sich aus der Wurzel hervorschießend ein grünes Gewächs, das eine Weile aus einem starken Stengel kräftige grüne Blätter nach den Seiten austreibt und zuletzt mit einer Blume endet.—Die Blume war unerwartet, überraschend, aber sie mußte kommen; ja das grüne Blätterwerk war nur für sie da und wäre ohne sie nicht der Mühe wert gewesen.] (FA, 2, 12: 209)

The flower in the analogy, Goethe explains, represents the ideal ending of the novel. In the "Novella," the lion and child demonstrate how love works better than force in containing the uncontainable: "This is the ideal—this is the flower" [Dies ist das Ideelle, dies die Blume] (FA, 2, 12: 210). Power is subdued not by force, but by its opposite: the loving attention of the child. The friendship between the lion and the child symbolizes an ideal reconciliation of heightened, polar opposites and is therefore an ideal reunion. This ideal reconciliation finds resonance in our higher nature (unsere höhere Natur) and is one "which proceeds from the heart of the poet" [das aus dem Herzen des Dichters hervorging] (FA, 2, 12: 210). A poet may reach the pinnacle of human nature if his art contains a meeting of intensified polar opposites. Love in the plant world is represented by the meeting of two intensified, articulated halves that unite physically and spiritually. The foliage of the plant is a necessary step toward intensification, and Goethe equates this stage to the prosaic world. To rise above the prosaic, Goethe, the poet, strives to imitate the process of plant metamorphosis. His poem stresses the reconciliation of opposed forces that serves as the climax of the whole. Nature's heightened stance in the flower, then, is one model for human creativity.

The foregoing examples, whether society intensifies to bring about a remarkable poet or whether a particular work of art culminates in the reconciliation of opposites, illustrates how closely Goethe believed human beings and their endeavors were linked with nature. He anthropomorphizes nature, but he also "naturalizes" human beings. In his view, the formation of specialized organs of a plant or the formation of red arises from a similar impulse as the one in human beings to create.

Before turning to the next chapter, which further explores the ability of nature and human beings to create entirely new entities, it is important to differentiate Goethe's notion of Steigerung from the more traditional one of teleology. Not only does Goethe's term assume that human beings are an integral part of nature (and do not stand above or apart from it), but it also places much more emphasis upon nature's creative powers.

The Case against Teleology

On the surface, it may appear that Goethe's principle of Steigerung, with its sense of increasing complexity or spirituality, implies a tacit acceptance of teleology—of a set hierarchy that was used from the time of Aristotle as an argument for patriarchal political structures, strict organic and hierarchical ends, and, ultimately, the contemplative life. Actually, however, Goethe distanced himself from teleology throughout his life.[33] While he clearly views some entities as higher, more complex, and more interesting than others, his principle of Steigerung allows for a great deal of flexibility and change. Steigerung enables the researcher to track the vitality of the organism without recourse to some cosmic arrangements of ends or some prearranged system within nature. By rejecting such an understanding of teleology, Goethe focuses his study on the organism itself—how it and its various parts create its own ends. Steigerung demonstrates that he is neither an ancient who believes in a set natural law, nor is he a modern who advocates complete relativity or equality of values. Rather, he combines aspects of both perspectives. Nature, in Goethe's view, contains too great a creative drive or desire to be held to predetermined ends.

Goethe's principle differs from Aristotle's concept of teleology in its emphasis on the great mutability of the final ends. For Aristotle, all living beings strive to participate in the immortal and the divine, but Aristotle limits the participation of nonhuman life to reproduction (*De Anima,* 415a27–b3). Moreover, Aristotelian teleology presents a strict hierarchy where the highest human acts are determined and ruled by reason.

Steigerung offers an alternative to the classical understanding of telos on several different levels. First, for Goethe, all natural things—human and nonhuman, organic and inorganic—may strive for creativity outside of procreation. Colors and plants both strive toward more complex manifestations. Second, the pinnacle or ultimate end of human happiness or flourishing is not philosophy or the contemplative life, but a creativity that combines the concerns of both matter and spirit. Third, where the traditional understanding of teleology sees the ends as permanent and eternal, Goethe's ends are constantly changing, evolving, creating themselves anew.[34] Because Steigerung aims at creativity, the hierarchy that it represents is not static. The moment is never frozen in time, but at each "final" stage something new may be created.

Goethe most thoroughly outlines his objections to teleology in an unpublished essay written around 1794, "Toward a General Comparative Theory" ("Versuch einer allgemeinen Vergleichungslehre"). In language similar to his later criticisms of the Newtonian doctrine, he finds fault with teleology as a traditional concept of the world that "the great majority accepts and follows unconditionally" (12: 53) [der große Haufe sich ohne weitere Bedingung unterwirft und nachfolgt] (FA, 1, 24: 209). He explains that "the progress of natural philosophy has been obstructed for many centuries" first "by the conception that a living being is created for certain external purposes" (12: 53), and second that its form (Gestalt) is predetermined by an archetypal force (Urkraft).[35]

Goethe similarly complains of how the concept of teleology has held scientists back in his "Metamorphosis of Plants: Second Essay (or Attempt/Experiment)" ("Metamorphose der Pflanzen: Zweiter Versuch"):

here it is not a question of whether the concept of final causes is convenient, or even indispensable, to some people, or whether it may not have good and useful results when applied to the moral realm. Rather, it is a question of whether it is an aid or a deterrent to physiologists in their study of organized bodies. I make bold to assert that it does deter them, therefore avoid it myself and consider it my duty to warn others against it. (Mueller, 80)

[daß hier nicht die Frage sei, ob die Vorstellungsart, der Endzweck manchen Menschen bequem, ja unentbehrlich sei, ob sie nicht aufs Sittliche angewendet gute und nützliche Wirkungen haben könnte, sondern ob sie den Physiologen der organisierten Körper förderlich oder hinderlich sei? welches letztere ich mir zu behaupten getraue, und deswegen sie selbst zu meiden und andere davor zu warnen für Pflicht halte.] (FA, 1, 24: 154–55)

The break with teleological theories represents for Goethe the break from tradition—a tradition that was willing to adhere to false scientific doctrines because of a perceived societal good.

Goethe specifically outlines his objections to teleology in "Toward a General Comparative Theory." He attacks the two tenets of teleology on the grounds that they inaccurately portray the workings of nature. He argues against the principles that hold (1) that morphological formation results only from a predetermined archetypal drive (Urkraft) outside of the organism,[36] and (2) that one organism is created for the purpose of another or one organ within an organism is created for a specific purpose of the whole organism. To Goethe, the first principle denies nature any kind of a willful (willkürlich) purpose or freedom to work outside of a given rule. The second principle further discounts other sudden organic changes or reactions due to environment or accidental circumstances.

Both tenets, therefore, rule out a sudden change or adaptation, and they force nature to adhere strictly to a set rule. As a consequence, teleology depicts nature as more constrained and regular than Goethe believes it actually to be. His principle of Steigerung is an attempt at a formulation that avoids the shortcomings of both aspects of teleology. First, Steigerung arises within the organism itself; second, organisms through Steigerung may not only take outside elements into account, but they may even use these elements as an impetus to change and develop.

Goethe describes the universe in this essay as an interlocking whole, where the organic and inorganic as well as internal and external forces constantly vie with one another. He postulates that such things as air, water, heat, and cold have a great deal of influence in shaping an organism's form and its functions:

The statement "the fish exists for the water" seems to me to say far less than "The fish exists in the water and by means of the water." The latter expresses more clearly what is obscured in the former, i.e., the existence of a creature we call "fish" is only possible under the conditions of an element we call "water," so that the creature not only exists in that element, but may also evolve there. The same holds true of all other creatures. . . . And this is all the more natural because the outer element can shape the external form more easily than the internal form. We can see this most clearly in the various species of seal, where the exterior has grown quite fishlike even though the skeleton still retains all the features of a quadruped. (12: 55–56)

[Der Fisch ist für das Wasser da, scheint mir viel weniger zu sagen als: der Fisch ist in dem Wasser und durch das Wasser da; denn dieses letzte drückt viel deutlicher aus, was in dem erstern nur dunkel verborgen liegt, nämlich: die Existenz eines Geschöpfes das wir Fisch nennen, sei nur unter der Bedingung eines Elementes das wir Wasser nennen möglich, nicht allein um darin zu sein, sondern auch um darin zu werden. Eben dieses gilt von allen übrigen Geschöpfen. Dieses wäre also die erste und allgemeinste Betrachtung von innen nach außen und von außen nach innen. . . . Und was noch mehr aber natürlich ist weil das äußere Element, die äußere Gestalt eher nach sich, als die innere umbilden kann. Wir können dieses am besten bei den Robbenarten sehn deren Äußeres so viel von der Fischgestalt annimmt wenn ihr Skelett uns noch das vollkommene vierfüßige Tier darstellt.] (FA, I, 24: 212–13)

The next chapter discusses in greater detail how internal and external forces lead to a greater sense of freedom of development in the organic world. Goethe's criticisms already in this early essay, however, revolve around the inability of human beings to accept that each organism exists for itself ("Zweck seiner selbst," as Goethe characterizes it in another early essay [FA 1, 24: 234]). He blames human conceit (Eitelkeit) for the belief that we are nature's final purpose (letzter Endzweck). Because of this conceit, we view and value everything in nature only in terms of its usefulness to ourselves as opposed to trying to gain a more comprehensive view of the whole of nature:

He [a human being] further believes that everything that exists is there for him, is there only as a tool and aid to his own existence. It follows as a matter of course that when nature provides tools for him, its acts within an intention and purpose equal to his own in manufacturing them. (12: 54)

[Glaubt er [der Mensch] ferner daß alles was existiert um seinetwillen existiere, alles nur als Werkzeug als Hülfsmittel seines Daseins existiere, so folgt wie natürlich daraus: daß die Natur auch eben so absichtlich und zweckmäßig verfahren habe, ihm Werkzeuge zu verschaffen, wie er sie sich selbst verschafft.] (FA, 1, 24: 211)

In viewing nature according to teleological principles, we falsely anthropomorphize it. For nature, however, each part of creation is considered equal, be it the thistle that increases the toils of human beings or the wheat that feeds us (FA, 1, 24: 210–11). If scientists are to progress at all in their endeavors, they need to rise above the trivial concept (trivialen Begriff) of teleology and begin to view nature as an integrated whole (FA, 1, 24: 211–12).[37] One will only be able to understand creative nature (bildende Natur) if one imitates nature by treating all things as part of an integrated whole. All things are related, and nothing is created with a predetermined purpose.[38] Since no creature is to be viewed as existing solely for the sake of another, the entire universe of organisms is seen as an organic whole (als einen Zusammenhang von vielen Elementen). Once we reach this stage of viewing the world, "we will no longer think of connections and relationships in terms of purpose and intention" (12: 56) [wir werden uns gewöhnen Verhältnisse und Beziehungen, nicht als Bestimmungen und Zwecke anzusehen] (FA, 1, 24: 214), but rather view all of nature as one community of numerous parts. According to Goethe, human beings can only understand their own role in the world if they view themselves as part of a wider community. Every organism has its own role within the whole community, and one must be careful not to treat some as totally unimportant.

Goethe also objects to teleology on spiritual grounds. He realizes that teleology had had a profound moral and political use, and he recognizes how closely moral and natural philosophical teachings could be linked together. For centuries, people had turned to teleology to justify everything from religious doctrines to absolute monarchy. To attack teleology was to question God's purpose in nature and the very basis upon which political structures were built. Goethe, however, argues that to worship a false image of the natural order could not represent true devotion. Rather than denying a place for God, God's purpose, or determinism in nature, he argues that recognizing nature's complexities brings about religious feelings as well:

We show disrespect neither for the primal force of nature nor for the wisdom and power of a creator if we assume that the former acts indirectly, and that the latter acted indirectly at the beginning of all things. Is it not fitting that this great force should bring forth simple things in a simple way and complex things in a complex way? Do we disparage its power if we say it could not have brought forth fish without water, birds without air, other animals without earth, that this is just as inconceivable as the continued existence of these creatures without the conditions provided by each element? (12: 55)

[Wir treten also weder der Urkraft der Natur, noch der Weisheit und Macht eines Schöpfers zu nahe, wenn wir annehmen: daß diese mittelbar zu Werke gehe, jener mittelbar im Anfang der Dinge zu Werke gegangen sei. Ist es nicht dieser großen Kraft anständig, daß sie das Einfache einfach, das Zusammengesetzte zusammengesetzt hervorbringe? Treten wir ihrer Macht zu nahe, wenn wir behaupten; sie habe ohne Wasser keine Fische, ohne Luft keine Vögel, ohne Erde keine übrigen Tiere hervorbringen können, so wenig als sich die Geschöpfe ohne die Bedingung dieser Elemente existierend denken lassen.] (FA, I, 24: 213)

As in "Reunion," God is not viewed as an active part of nature per se but as the force which began a creative activity that continues to this day. One recognizes the creator more truly if one does not adopt a teleological stance because it would deny the creative urge to all of nature, a creative urge that, for Goethe, defines God. Similarly, here, too, Goethe asks, "But will we not show more regard for the primal force of nature, for the wisdom of the intelligent being usually presumed to underlie it, if we suppose that even its power is limited, and realize that its forms are created by something working from without as well as from within?" (12: 54) [Wird uns aber nicht schon die Urkraft der Natur die Weisheit eines denkenden Wesens welches wir derselben unterzulegen pflegen, respektabler, wenn wir selbst ihre Kraft bedingt annehmen, und einsehen lernen daß sie eben so gut von außen als nach außen, von innen als nach innen bildet] (FA, 1, 24: 212). Once again, Goethe's God is not the God of the Judeo-Christian world, but a limited being. He is a being, moreover, whose creative essence is mirrored in nature. Moreover, as I discuss in the next chapter, his image of a creative God fashioning new forms within limitations becomes the metaphor for nature's ability to evolve.

Goethe's aversion to teleology continued throughout his life. In "Experiment as Mediator" (1793), the essay that lays out his criteria for experiments, he argues that one "must remain unmoved by beauty or utility" [So soll . . . weder die Schönheit noch die Nutzbarkeit der Pflanzen rühren], and one should strive to "view nature's objects in their own right and in relation to one another" (12: 11) [die Gegenstände der Natur an sich selbst und in ihren Verhältnissen unter einander zu beobachten] (FA, 1, 25: 26). He explains in a later essay on meteorology in 1825 (FA, 1, 25: 275) that it is often so difficult to distinguish the cause from the effect that it is all too easy to confuse the one with the other. He also explained to Eckermann that, although he was unfamiliar with Kant's *Third Critique* at the time in which he wrote "Metamorphosis,"[39] this work is

wholly in the spirit of his doctrine. The separation of subject from object, and further, the opinion that each creature exists for its own sake, and that cork-trees do not grow merely that we may stop our bottles—this Kant shared with me, and I rejoiced to meet him on such grounds. (11 April 1827)

[ganz im Sinne seiner Lehre. Die Unterscheidung des Subjekts vom Objekt, und ferner die Ansicht, daß jedes Geschöpf um sein selbst willen existiert und nicht etwa der Korkbaum gewachsen ist, damit wir unsere Flaschen propfen können, dieses hatte Kant mit mir gemein und ich freute mich ihm hierin zu begegnen.] (FA, 2, 12: 243)

Goethe also shared with Kant a rejection of viewing nature exclusively in terms of mechanical principles and of attributing to nature's actions any divine designs.[40] For both men, biology had to be approached by treating the organism apart from mechanical principles because an organism acted on some level as its own agent. Both men, therefore, emphasized the organism's self-determination (Erpenbeck, "Wissenschaft," 1189). In his *Third Critique,* Kant, like Goethe was battling the propensity of the time to treat Newtonian mechanistic categories as "absolute realities." As Cornell explains, "The various ways in which life was conceived in this period—according to extrinsic or intrinsic vital powers, incorporeal or material agencies, or a distinct set of created laws—all recall the metaphysical abstractions Newton employed and suggest their dogmatic adoption, often supported by a doctrine of God's creation through laws and forces" ("Newton," 409).

Late in his life, Goethe was still quite critical of teleology. In 1830, he praised Geoffroy St. Hillaire's theories for clearing away "the sad expedient of final causes" (den traurigen Behelf der Endursachen [FA 1, 24: 836]). In a review (1831) of Vaucher's work, Goethe further reiterates his own aversion to teleological views (teleologische Ansichte), "which are not and could never be ours" (Mueller, 211) [welche die unsrigen nicht sind noch sein können] (FA 1, 24: 773).

We can further see how pervasive Goethe's rejection of teleology was by turning to a discussion on art and collecting. In speaking of museums and art galleries in his essay on Winckelmann (1805), Goethe remarks,

It is sad to have to think of anything as final and complete. Old armories, galleries and museums to which nothing is added are like mausoleums haunted by ghosts. Such a limited circle of art limits our thinking. We get accustomed to regarding such collections as complete, instead of being reminded through ever new additions that in art, as in life, we have nothing that remains finished and at rest, but rather something infinite in constant motion. (3: 113)

[Traurig ist es, wenn man das Vorhandne als fertig und abgeschlossen ansehen muß. Rüstkammern, Galerien und Museen, zu denen nichts hinzugefügt wird, haben etwas Grab- und Gespensterartiges; man beschränkt seinen Sinn in einem so beschränkten Kunstkreis, man gewöhnt sich, solche Sammlungen als ein Ganzes anzusehen, anstatt daß man durch immer neuen Zuwachs erinnert werden sollte, daß in der Kunst, wie im Leben, kein Abgeschlossenes beharre, sondern ein Unendliches in Bewegung sei.] (FA, 1, 19: 198)

All things, including such inanimate entities as collections, ought never to be viewed as complete, but always in progress. Only that which is able continually to change is able to expand our way of thinking. Once something is considered complete, it becomes dead and is not able to move us. Goethe's warning to art collectors mirrors his warning to scientists: if one assumes an end product, one limits one's ability to understand the object of one's study. For scientists, Steigerung involves striving toward knowledge of nature, but because nature is not fixed in the Platonic sense of an ideal, scientists must constantly revisit the goals of their research. Moreover, since scientists are imitating nature in their study of it, the scientific model should not be static. Rather, scientists must constantly be creative themselves in their approach to nature if they are to understand it.

The issues involved in differentiating Steigerung from teleology are not that different from current literary and scientific debates. On the one hand, Goethe, like many scientists and humanists of our own time, believes that nature is capable of randomness and spontaneity, and is not simply a determined force. On the other hand, however, he portrays a kind of upward-striving, creative will in nature. This striving, albeit not static and determined, nevertheless allows for a kind of pattern that guides progressive development. Perhaps Goethe's middle position is not all that different or, for that matter, unrelated to current discussions in science.[41] Physicists, even in light of chaos theory and quantum physics, are still debating whether nature follows patterns of behavior. Some, for example, argue that quantum physics and chaos theory do not preclude determinism in nature.[42] Nature, therefore, is viewed as chaotic, yet patterned.

❧

While creativity has been a source of much of the foregoing discussion, the link is still missing between Steigerung and aesthetic perfection. How is it that in Goethe's view intensification leads to something more beautiful and more perfect than things that have not undergone intensification? If intensification is never really separable from polarity, how is it that these two opposing forces create something beautiful? How, according to Goethe's theory, are we supposed to recognize beauty that is modeled upon nature? To answer these questions, it becomes necessary to turn to Goethe's principle of compensation, a principle that describes how the interaction of polarity and intensification, freedom and limits, and internal and external drives leads to beauty.

3 THE PRINCIPLE OF COMPENSATION AND THE CREATION OF NEW FORMS

IN *Elective Affinities,* Goethe's most scientific novel, he somewhat strangely draws our attention to the characters' relationships with money: the rich Baron Eduard spends so freely that he nearly bankrupts his estate; Charlotte, his wife, constantly tries to control the purse strings; their friend, the Captain, despite his numerous talents, will not take a job and instead relies upon others for sustenance; and Ottilie, Charlotte's niece, is so self-denying that she refuses to touch the money given to her and changes clothing only for reasons of cleanliness.[1] The budget of each of the characters serves as a metaphor for their main traits and offers us a tool with which to analyze their personalities. Eduard is enormously generous, but he also refuses to deny himself anything, no matter what the consequences. Charlotte tries to maintain moral and financial order, but she is so controlling that she eventually feels responsible for her child's death. The Captain tries to be tremendously helpful on the estate, but he never breaks free from his dependence on others, either emotionally or financially. And Ottilie, like the Captain, tries to be as helpful and as unobtrusive as she can. These desires, however, lead to her suicide.

In Goethe's scientific works, he similarly turns to a budget metaphor, the principle of compensation, to track the particular strengths and weaknesses of organic and inorganic forms. He tells us, for example, that if we see an animal that is particularly advantaged in one way, we ought immediately to search for its complementary deficit or disadvantage. This principle involves, for Goethe, one of nature's most sophisticated means of creativity. Each organism's "budget" (its size, physical makeup, environment, etc.) symbolizes the limitations with which it is born. That new forms constantly evolve illustrates nature's ability to create a myriad of forms within severe restrictions. Nature's urge to be creative at times transcends the limitations presented to it. This great malleability of nature, however, also represents one of science's greatest challenges: how is one to study a form that is constantly in flux?

Goethe's theory of compensation has been misunderstood in light of its connection to his idea of the animal type (Typus). Several scholars have turned to his use of a type to argue that he was an essentialist, i.e., that he believed in the existence of unchanging natural types similar to Platonic Ideals (Mayr, 457–48; Wells, 29–30). For example, scholars often mistakenly link his idea of the type with an interpretation of his "Metamorphosis" essay that emphasizes the leaf's underlying *eidos* (Mayr, 457–48). Thus, instead of focusing on the type's—or leaf's—malleability, they see it as support for a static type.[2] As a result, they conclude that Goethe's "ideas had nothing to do with evolution," but at best were "vague anticipations of principles later formulated by Geoffroy" (Mayr, 458). A close examination of Goethe's use of the type, however, suggests that he did not believe in the existence of a type.[3] To the contrary, he specifically emphasizes that it is an artificial construct that he invents to enable the study of the *fluidity* of nature. In other words, the type does not demonstrate that Goethe's ideas could not be evolutionary because he focused on unchanging forms. Rather, it emphasizes Goethe's belief in the variability of nature and thus, at the very least, leaves the door open to the question of Goethe's evolutionary beliefs. This door, moreover, is further opened by Goethe's account of changing forms, his comparisons of human beings with different animals and, perhaps most significantly, his belief in the variability of individuals.

Descartes starkly differentiates animals from human beings. Animals, according to his philosophy, have no souls, but are complicated machines whose parts will one day be completely reproducible through science. Similarly, Newton presents mathematical formulas for categorizing and defining the world. He believes that nature operates according to predictable mathematical equations. Nature could and should be compartmentalized and its parts separately analyzed. Goethe certainly was not alone in rejecting Cartesian and Newtonian biology, and many of his contemporaries (e.g., Schelling and Oken) went even further than Goethe in portraying a dynamic view of nature (Jardine, 233). However, while many eighteenth-century scientists saw nature as more vitalistic than did either Newton or Descartes, many still saw nature, apart from human nature, as primarily reactive and explicable through natural laws. The most prominent and influential naturalists of Goethe's time were operating under the rubric of Enlightenment science, and Goethe's morphological studies seek to refute their methodologies. Buffon (1707–88), especially in his early writings, "was rather a strict Newtonian" and "was impressed by concepts of movement and continuity" (Mayr, 331). Mayr even links the inability of Buffon and Cuvier (1769–1832) to see the link between human beings and animals as an inherently Cartesian problem: both men accepted the Cartesian claim that animals were so qualitatively different from human beings that it was quite "impossible to consider man as having evolved from animals" (332, 370). As far as evolution is concerned, the first "definite proposal" of evolution is generally considered to have come from Lamarck (1744–1829), who had a "profound intellectual commitment to Descartes, Newton, Leibniz, and Buffon" (Mayr, 342–44). Lamarck, in his view of evolution, was not a vitalist, but accepted "only mechanistic causes." He adopted from Newton "a belief in the lawfulness of the universe and the conviction that all phenomena . . . could be explained in terms of movements and forces acting upon matter" (Mayr, 357, 344).[4] Like most of his contemporaries (e.g., Buffon, Blumenbach, Forster, and d'Alton), Lamarck attributed gradual changes in animal forms to a reaction to changes in their environment.[5]

Even those naturalists who were more or less adherents of Goethean science attributed less creativity to nature's organisms than did Goethe. While several of the Romantic scientists recognized nature's ability to change certain external forms, "the inner being of nature" was "supposed to remain untouched by this transformation" (Engelhardt, 58). Thus, Schelling and Hegel could not have thought of descent, according to Engelhardt, because they viewed certain aspects of nature as unchanging (58–59). Even d'Alton, who acknowledges Goethe's influence, almost completely discounts animal creativity. For d'Alton, there is no great difference between a beaver's dwelling and a snail's shell. Both arise out of an animal's dumb instinct for shelter (*Nagethiere*, 1–2). Nor did the vitalists, in the end, exert the most influence. The mechanistic argument to explain life and reproduction continued into the nineteenth and twentieth centuries. Darwin and his followers are generally regarded as Newtonians in their worldview.[6] For Ernst Haeckel (1834–1919), "Darwin was the 'new Newton' who had explained organisms strictly by mechanical causes—invalidating Kant's claim that not even a 'blade of grass' would be accounted for without the principle of purposiveness" (Cornell, "Newton," 405). Nietzsche, too, labeled Darwin as a mechanist (315). In his recent book, Lewontin argues that evolutionary biologists still follow the Cartesian metaphor of the world as a machine—a metaphor that has significantly fashioned their mode of inquiry (3–4, 38, 71–73).

One of the key scientific debates of the nineteenth century focused upon the issue of organic forms and pitted a more mechanical understanding of the world against a more vitalistic one. In 1830, Georges Cuvier debated Étienne Geoffroy Saint-Hilaire (1772–1844) over the causes behind particular animal features.[7] Cuvier "championed an extreme empirical methodology," whereas "facts" for Geoffroy "were only the building blocks of science; the essence of science was ideas" (Appel, 6–7). Cuvier promoted a method of studying animal forms by analyzing disparate parts, whereas Geoffroy presented a doctrine that examined the interrelationship of parts.[8] Cuvier believed that teleology explained the formation of particular parts, whereas Geoffroy believed that changes in structure "will cause changes in function" (Mayr, 463).

Goethe avidly followed the debates and sided with Geoffroy. A piece of anecdotal evidence illustrates how important this debate was for Goethe.[9] On 2 August 1830, Soret reports how the news of the July Revolution had reached Weimar and caused quite a commotion. Soret went to see Goethe, who exclaimed, "Now . . . what do you think of this great event? The volcano has come to an eruption; everything is in flames, and we have no longer an affair behind closed doors" [Nun . . . was denken Sie von dieser großen Begenbenheit? Der Vulkan ist zum Ausbruch gekommen; Alles steht in Flammen, und es ist nicht ferner eine Verhandlung bei geschlossenen Türen] (FA, 2, 12: 726). Soret mistakenly believed that Goethe was speaking of the political revolution. Instead, Goethe was speaking of the "important contest" for science: the debates between Cuvier and Geoffroy. Soret was not unexpectedly left speechless by Goethe's comments. Goethe, however, explains in great detail why this debate was so important to him. It signaled for him not only an important scientific debate, but a piece of evidence that his kind of scientific approach was gaining ground over a Newtonian one:

From the present time, mind/spirit will rule over matter in the physical investigations of the French. There will be glances of the great maxims of creation, of the mysterious workshop of God! Besides, what is all intercourse with nature, if, by the analytical method, we merely occupy ourselves with individual material parts, and do not feel the breath of the spirit, which prescribes to every part its direction, and orders, or sanctions, every deviation, by means of an inherent law?

I have exerted myself in this great affair for fifty years. At first, I was alone, then I found support, and now at last, to my great joy, I am surpassed by congenial minds. When I sent my first discovery of the intermaxillary bone to Peter Camper, I was, to my infinite mortification, utterly ignored. With Blumenbach I fared not better, though, after personal intercourse, he came over to my side. But then I gained kindred spirits in Sömmering, Oken, d'Alton, Carus, and other equally excellent men. And now Geoffroy Saint-Hilaire is decidedly on our side, and with him all his important scholars and adherents in France. This occurrence is of incredible value to me; and I justly rejoice that I have at last witnessed the universal victory of a subject to which I have devoted my life, and which, moreover, is my own par excellence. (*Oxenford translation)

[Von nun an wird auch in Frankreich bei der Naturforschung der Geist herrschen und über die Materie Herr sein. Man wird Blicke in große Schöpfungsmaximen tun, in die geheimnisvolle Werkstatt Gottes!—Was ist auch im Grunde aller Verkehr mit der Natur, wenn wir auf analytischem Wege bloß mit einzelnen materiellen Teilen uns zu schaffen machen, und wir nicht das Atmen des Geistes empfinden, der jedem Teile die Richtung vorschreibt und jede Ausschweifung durch ein inwohnendes Gesetz bändigt oder sanktioniert!]

[Ich habe mich seit fünfzig Jahren in dieser großen Angelegenheit abgemüht; anfänglich einsam, dann unterstützt, und zuletzt zu meiner großen Freude überragt durch verwandte Geister. Als ich mein erstes Aperçu vom Zwischenknochen an Peter Camper schickte, ward ich zu meiner innigsten Betrübnis völlig ignoriert. Mit Blumenbach ging es mir nicht besser, obgleich er, nach persönlichem Verkehr, auf meine Seite trat. Dann aber gewann ich Gleichgesinnte an Sömmering, Oken, d'Alton, Carus und anderen gleich trefflichen Männern. Jetzt ist nun auch Geoffroy de Saint-Hilaire entschieden auf unserer Seite und mit ihm alle seine bedeutenden Schüler und Anhänger Frankreichs. Dieses Ereignis ist für mich von ganz unglaublichem Wert, und ich jubele mit Recht über den endlich erlebten allgemeinen Sieg einer Sache, der ich mein Leben gewidmet habe und die ganz vorzüglich auch die meinige ist.] (FA, 2, 12: 727–28)

I have quoted Goethe's comments in detail because they reveal the extent to which he held Newtonian and Cartesian approaches influenced the natural sciences. I do not by any means interpret Goethe's enthusiasm in this context as primarily a support for an evolutionary theory, but rather a more general condemnation of materialist or mechanical explanation for organic events: the subject to which he devoted his life is the limiting quality of Newtonian approaches. His reference to the triumph of the spirit in the first paragraph above refers to a vitalistic and will-driven component within nature that is present in so many of his scientific writings. Changes in nature do not occur solely due to mechanical reactions to causes as the Newtonian-influenced naturalists argued, but may also come about due to an interior quality of nature: an impulse, an expression of will from within that may lead to organic change. Goethe is happy with Geoffroy's methodology because it, like Goethe's own, views the whole organism within an interrelated context of both its parts and the environment. Like Goethe, Geoffroy believed in the law of compensation[10] and, like Geoffroy, Goethe emphasized synthesis over analysis (FA, 1, 24: 822). And although not the primary focus of the foregoing quotation, Goethe's reference to the intermaxillary bone certainly suggests that he viewed the Cuvier-Geoffroy debates as relating to the greater question of the relationship between human beings and other animals.

This quotation only intimates some of the larger issues of organic changes and transformations within Goethe's scientific works. A study of his works further outlines the importance of the spirit (Geist) of nature—especially in the sense that the spirit of one individual may suddenly bring about an evolutionary change. Goethe's animals have a fierce desire not simply to exist but to flourish. Like Faust, some animals, plants, and even inanimate entities display an extraordinary will or desire to strive toward self-overcoming. This drive leads to creativity, whether in the form of a beaver's dam or the creation of an entirely new animal form.

According to Goethe's morphological texts, some entities in nature are driven or inspired from within to create something superfluous or something entirely new. This inner drive, moreover, is something quite different than a Darwinian survival of the fittest or a reactive adaptation of the environment. Changes in organic form in Goethe's account do not arise solely because of external causes. Rather, Goethe credits certain organisms, and even parts of organisms, with a desire to express their own will: changes in form are not inexplicable (spontaneous/random), nor are they simply reactive. Organisms choose to change environments, strive toward beauty of form, and adapt to obstacles. In his scientific writings, he will describe plants that strive toward beauty of form apart from reproductory concerns, or a creature of the sea that suddenly takes it upon itself to live a more difficult life on land. Although nature often follows regular principles, at any time any organism or phenomenon may exert a capricious will and break free of its regular principles (GA, 17: 710–11). Nature is also at its most creative at precisely those moments of deviation (FA, 1, 24: 462). New organic formations and new natural phenomena arise during moments of transition and deviation from regular patterns.[11] Because nature is creative even within its most basic principle, Goethe praises it as an artist and model for human art (e.g., FA, 1, 24: 415; FA, 1, 25: 63; FA, 1, 2: 838–39) and views it as an ever-changing Proteus (FA, 1, 24: 93, 234). While scholars have discussed whether and to what extent Goethe believed in descent, the more central question for Goethe's science may well not be how the process of evolution occurs, but why it occurs.

The applications of the principle of compensation, the principle behind Goethe's theory of organic change, are widespread throughout his corpus. It plays a significant role in his descriptions of the natural world, his scientific methodology, his analysis of human character, and his understanding of beauty. He employs this principle to explain animal development, the success rates of certain species, upright posture, cloud formations, and even psychological imbalances. Chapter 1 investigated some of the aesthetic consequences of Newtonian optics. They deaden our appreciation for the dramatic and colorful theater of nature. Goethe's principle of compensation even further assails modern science. Without an understanding of this principle, he argues that human beings will never be able to recognize, foster, and create truly beautiful things or discover the relationship between natural beauty and human freedom.

Scientific Methodology and the Type

Goethe most thoroughly describes his compensation principle in one of his earlier scientific essays, "Outline for a General Introduction to Comparative Anatomy, Commencing with Osteology" ("Erster Entwurf einer allgemeinen Einleitung in die vergleichende Anatomie, ausgehend von der Osteologie"). This essay (written in 1795 and published in 1820) advocates establishing a new method for comparing animal species.[12] He begins with the problem of methodology before addressing the way in which animal forms themselves change. He notes that, in the past, it has been extraordinarily difficult to study animal forms because of their ability to transform over time. Because animal forms undergo constant flux, it is first necessary to develop a proper methodology to follow na-ture's changing forms. Therefore, he addresses the issue of methodology before proceeding with the actual study of the forms. He explains that one of the difficulties in conducting comparative anatomy has been the lack of a standard for comparison. Too many different perspectives and too many different animal forms were jumbled together. This science of anatomy therefore had no limits (keine Grenzen) because its field of in-quiry was too large. As a consequence, scientific research suffered as "any empirical investigation exhausts itself in a vast extent" (12: 117) [jede bloß empirische Behandlung müdet sich ab in dem weiten Umfang] (FA, 1, 24: 228). He mentions three kinds of inadequate approaches due to the lack of limits in this scientific field: (1) a too general approach that never was able to transcend simple appearances, (2) a teleological method that did not focus enough upon the individual organism, and (3) a pious approach that sought to glorify God but became lost in empty speculation (leere Spekulation) (FA, 1, 24: 228).

Just as Goethe complains in his "Metamorphosis" that Linnaeus was incapable of progressing due to a superfluity of data, so too in compara-tive anatomy were scientists hindered from progression due to the broad expansiveness of their field of study. To solve this problem, Goethe sug-gests comparing various animals not against one another, but against an artificially created, general animal type (Typus). This type would be a kind of comparative canon (Vergleichungskanon) because it serves as a somewhat stable measure against which one may either compare several different animals or mark the changes within one species. This type would refer to all animals and not to a specific animal. Much of the con-fusion over the role of the type has arisen due to Goethe's Platonic ter-minology. He refers to the type as the idea (Idee) that governs the whole (FA, 1, 24: 229). The purpose of the type, however, is not to discover an underlying similarity, but to enable the study of fluid forms.

The type becomes a temporary means of discussing a particular stage in nature's development. It acknowledges nature's great malleability but also simultaneously becomes a limit on the data for comparison. More-over, where the doctrine of Platonic forms teaches that if we study vari-ous forms and phenomena, we will be able to identify an underlying *simi-larity* of form, Goethe uses his type to discover and highlight the *differences* between individuals as well as species. In opposition to the mathematical formulas of Newton and the mechanistic philosophy of Descartes, Goethe reminds scientists that they must be willing to change their constructs as nature itself changes. The type is not to become permanent. Nature cannot be reduced to the simplest terms of algebra or plotted onto a graph because it is capricious and ever-changing. In an introductory essay to his morphology journal (1817), he instructs scientists to be as fluid as nature itself:

When something has acquired a form it metamorphoses immediately to a new one. If we wish to arrive at some living perception of nature we ourselves must remain as quick and flexible as nature and follow the examples it gives. (*12: 64)

[Das Gebildete wird sogleich wieder umgebildet, und wir haben uns, wenn wir einigermaßen zum lebendigen Anschaun der Natur gelangen wollen, selbst so beweglich und bildsam zu erhalten, nach dem Beispiele mit dem sie uns vorgeht.] ("Die Absicht eingeleitet," FA, I, 24: 392)

Goethe further warns that once researchers adopt a particular method, it becomes very easy to forget that they have imposed part of themselves upon their natural studies. He writes that natural observers must frequently remind themselves that created types or groupings are artificial and not natural. Scientists must approach their research with a self-consciousness capable of questioning the reasons behind their particular models. To distinguish between human systems and natural principles, he exhorts scientists in another work ("Problem und Erwiderung," 1823) always to return to close observations of nature as a litmus test of their method: "But our full attention must be focused on the task of listening to nature to overhear the secret of her process, so that we neither frighten her off with coercive imperatives, nor allow her whims to divert us from our goal (12: 44) [Unsere ganze Aufmerksamkeit muß aber darauf gerichtet sein, der Natur ihr Verfahren abzulauschen, damit wir sie durch zwängende Vorschriften nicht widerspenstig machen, aber uns dagegen auch durch ihre Willkür nicht vom Zweck entfernen lassen] (FA, I, 24: 584). His method of temporarily fixing nature or comparing it against an artificial type is an attempt to limit nature enough to allow an investigation of it, while fluid enough to observe an ever-changing nature.

Goethe explains in his "Comparative Anatomy" essay that by using a general type, scientists and other experts (horsemen, butchers, etc.)

from different fields may agree on a common basis for comparison and may therefore limit themselves to the same starting point (Vereinigungspunkt).[13] It would thus eliminate the problems associated with earlier attempts to study comparative anatomy. Specifically, one might compare more complicated aspects of organisms and their parts and therefore be able better to observe nature's malleability of forms; one would see more clearly the complicated nexus of internal and external relationships of the parts and therefore abandon simplistic teleological accounts for more dynamic ones; and, perhaps most importantly, one would more freely compare human beings with other animals and therefore see the continuity of nature in all organic forms.

Goethe believes that his comparative type will illustrate the unity of nature and the place of human beings within it (FA, 1, 24: 229). Although he admits that our animal characteristics have been intensified for higher purposes [Im Menschen ist das Tierische zu höhern Zwecken gesteigert] (FA, 1, 24: 228), human beings are still governed by the same principles as the rest of nature. To see our place within nature, we are to begin with more simple forms and then ascend (steigen) to human beings from the other animals (FA, 1, 24: 237). He implies a kind of evolutionary development, especially when he argues that his readers will be better able to understand certain aspects of the human form and structure once they have studied lower animal forms (FA, 1, 24: 237; cf. FA, 1, 24: 265ff.). He further notes that scholars denied the existence of the intermaxillary bone because "this trait was to distinguish us from the apes" (12: 124) [hier sollte das Unterscheidungszeichen zwischen uns und dem Affen sein] (FA, 1, 24: 239). Goethe, who was one of the first to argue for the existence of this bone in the human skull, laments that such a denial (by Camper and Blumenbach among others) hindered the development of comparative anatomy (FA, 1, 24: 239).[14]

Although Goethe's type seems to endorse a kind of fixed standard with which to measure and understand nature, he is adamant that this standard must be flexible. He, in this sense, was neither an idealist nor a strict essentialist. Establishing a type was intended as a means to trace particular aspects of individual animals and species and not to fix or idealize the forms themselves. Unlike the Platonic Ideal, this type itself is subject to change. It is not to become a fixed means of studying nature; scientists must constantly reformulate the type as they test their various hypotheses (FA, I, 24: 230). Goethe justifies this semi-static type because organic forms change so frequently that scientists might otherwise lose sight of them. As he would later write (1817), a close examination of nature and its forms reveals "that nothing in them is permanent, nothing is at rest or defined—everything is in a flux of continual motion" ("The Purpose Set Forth," 12: 63) [daß nirgend ein Bestehendes, nirgend ein Ruhendes, ein Abgeschlossenes vorkommt, sondern daß vielmehr alles in einer steten Bewegung schwanke] (FA, I, 24: 392). Thus, the scientist, like nature itself, must be willing to be versatile and see the various different deviations from the type, if he or she is to grasp nature in all of its complexity:

But now that we have endured in the realm of what is enduring, we must also learn to change our views along with ever-changing nature. We must learn many different movements so that we grow deft enough to follow the type in all its versatility, and so that this Proteus never slips from our grasp. (*12: 121)

[Nun aber müssen wir, indem wir bei und mit dem Beharrlichen beharren, auch zugleich mit und neben dem Veränderlichen unsere Ansichten zu verändern und mannigfaltige Beweglichkeit lernen, damit wir den Typus in aller seiner Versatilität zu verfolgen gewandt seien und uns dieser Proteus nirgend hin entschlüpfe.] (FA, I, 24: 234)

Because nature is both fixed and free, enduring and changing, Goethe advises scientists to adopt methods that can track and further mirror these polar aspects. He insists that the type cannot be rooted in permanence, but that it must be subject to change.

The Principle of Compensation

After Goethe has discussed the creation of the type, he begins to focus upon the principle of compensation, the principle behind nature's different and ever-changing forms. Like Aristotle, who was one of the first to write about the principle of compensation,[15] Goethe suggests that animals must balance each of their features against others. For example, animals with tusks have expended so much material on this feature, that they do not have enough material left over for horns on top of their heads.[16] Unlike Aristotle, Goethe believes that the principle of compensation explains not only particular features within one species, but also how animals change and evolve over time and how they create new forms. If nature did not have so much room to play (Spielraum) within its own rules, he explains, it could never produce so many endless forms from this simple budget or image [Bild] (FA, 1, 24: 244). Nature is free (frei) to allocate the resources within the limitations of the budget. Within the framework of the budget, organisms may carve out new forms for themselves.

When Goethe discusses nature's almost limitless ability to create new forms, he uses the language of bookkeeping to describe the creative processes of organic development and variations. He describes animal formation in terms of dynamic, economic trade-offs (Etats, Budget, haushälterisches Geben und Nehmen). He gradually replaces the idea of an animal type with the notion of a balance sheet or budget, where every feature has a price, whether it is the extraordinarily large uterus of a hen or the lion's noble mane. The "cost" of each such feature must be deducted from the budget, leaving less available funds for other parts. According to Goethe's plan, if one wishes to study specific animal species, one must consider how they deviate from the general budget:

Thus, for example, the neck and extremities are favored in the giraffe at the expense of the body, but the reverse is the case of the mole.

In the above observation we encounter this law: nothing can be added to one part without subtracting it from another, and vice versa. (12: 121)

[So sind, zum Beispiel, Hals und Extremitäten auf Kosten des Körpers bei der Giraffe begünstigt, dahingegen beim Maulwurf das Umgekehrte statt findet.]

[Bei dieser Betrachtung tritt uns nun gleich das Gesetz entgegen: daß keinem Teil etwas zugelegt werden könne, ohne daß einem andern dagegen etwas abgezogen werde, und umgekehrt.] (FA, 1, 24: 233)[17]

One can see how polarity here, too, plays a role. Nature is faced with contrary pulls as its forms develop. However, Goethe's theory of compensation adds a new element to his most basic principle: that of limits. The limits placed upon nature play an important role in the future development of organic as well as inorganic forms. In addition, one may also see the role of striving within this principle. Nature's various forms try to exert a kind of willful defiance against the limits imposed upon them, and it is the interplay of polarities, striving, and limits that eventually leads to the creation of new forms.

Goethe expresses a similar idea within his poem "Metamorphosis of Animals" ("Metamorphose der Tiere"). Its central idea is the principle of compensation, which operates as the key (Schlüssel) to all formation. A large part of animal development in the poem, as throughout his scientific texts, resides within the role of limits. The principle of compensation, the highest law (das höchste Gesetz), highlights those limits. Nature gives each organism a certain budget, and even God does not broaden the limits placed by nature because "Only a limit enables a form to rise to perfection" [Denn nur also beschränkt war je das Vollkommene möglich] (FA, 1, 2: 499). When the poet lists some of the main limits upon animal formation, they appear to be circular. The animal's shape determines its way of life, while its way of life determines its shape:

> So the shape of the animal patterns its manner of living,
> Likewise their manner of living, again, exerts on the animals'
> Shapes a massive effect . . .
> (1: 161)

> [Also bestimmt die Gestalt die Lebensweise des Tieres,
> Und die Weise zu leben, sie wirkt auf alle Gestalten
> Mächtig zurück . . .]
> (FA, 1, 2: 499)

As in the essays on comparative anatomy, the animal's ultimate shape and appearance are influenced by external and internal forces. Goethe's rejection of the role of teleology in the formation process is equally apparent: "Every animal *is* an end in itself . . ." (1: 161, original emphasis) [Zweck sein selbst ist jegliches Tier . . .] (FA, 1, 2: 499). Animals are not created by nature for particular purposes nor are their individual parts pre-formed.

At first, the budget theory, or the principle of compensation, seems to emphasize the limitations of nature. He admits that animal forms are greatly influenced by outside circumstances and informs us that our first task is to investigate the outer influences upon organisms. Creatures living in the water, for example, will emerge with completely different features than creatures living upon the land. Water will have a bloating effect upon the bodies of organisms that live in the water, and hence one will find a contraction of extremities and outer organs (FA, 1, 24: 235).[18]

The true force of this theory is that animals create the most varied forms for themselves within the scope of the very limits they face. In the "Polarity" essay, Goethe praises nature for creating a multiplicity of forms within the scope of defined limits. In the "Comparative Anatomy" essay, he uses economic language to make a similar point:

These are the bounds of animal nature; within these bounds the formative force seems to act in the most wonderful, almost capricious way, but is never able to break out of the circle or leap over it. The formative impulse is given hegemony over a limited but well-supplied kingdom. Governing principles have been laid down for the realm where this impulse will distribute its riches, but to a certain extent it is free to give to each what it will. If it wants to let one have more, it may do so, but not without taking from another. Thus nature can never fall into debt, much less go bankrupt. (12: 121)

[Hier sind die Schranken der tierischen Natur, in welchen sich die bildende Kraft auf die wunderbarste und beinahe auf die willkürlichste Weise zu bewegen scheint, ohne daß sie im mindesten fähig wäre den Kreis zu durchbrechen oder ihn zu überspringen. Der Bildungstrieb ist hier in einem zwar beschränkten, aber doch wohl eingerichteten Reiche zum Beherrscher gesetzt. Die Rubriken seines Etats, in welche sein Aufwand zu verteilen ist, sind ihm vorgeschrieben, was er auf jedes wenden will, steht ihm, bis auf einen gewissen Grad, frei. Will er der einen mehr zuwenden, so ist er nicht ganz gehindert, allein er ist genötigt an einer andern sogleich etwas fehlen zu lassen; und so kann die Natur sich niemals verschulden, oder wohl gar bankrott werden.] (FA, 1, 24: 233–34)

Although the organism can never spend beyond its "bank account," it has a great deal of freedom to determine how to spend the "funds" given to it. It has a certain autonomy in determining which features and qualities it wishes to emphasize: "we will find that a limit is set to nature's structural range, but the number of parts and their modifications allow for the form to be changed ad infinitum" (12: 121) [so finden wir, daß der Bildungskreis der Natur zwar eingeschränkt ist, dabei jedoch, wegen der Menge der Teile und wegen der vielfachen Modifikabilität, die Veränderungen der Gestalt ins Unendliche möglich werden] (FA, 1, 24: 233). For example, if one compares the snake, the lizard, and the frog, one can see how a very similar form through slight variations leads to the creation of different animals (FA, 1, 24: 235).

For Goethe, each animal is its own small world (eine kleine Welt). He here reiterates his objections to teleology by explaining that each animal is a purpose unto itself. Every part and organ has a relationship to the whole. A predominance in one part will cause a decline in another. The organism arises not in isolation, but in a dynamic relation of its own inner drive together with environmental influences. Goethe's objections to analysis seem to stem from his view of organic development. If one takes the animal apart or examines it outside of the context of its other parts, one already has discounted one of the major influences upon its development. In particular, one will never understand what has caused the organism to favor one feature over another. He then turns the argument of teleology upside down:

Thus, in the future, members such as the canine teeth of the *Sus babirussa* will not elicit the question, What are they for? but rather, Where do they come from? We will not claim that a bull has been given horns so that he can butt; instead, we will try to discover how he might have developed the horns he uses for butting. (12: 121)

[Man wird also künftig von solchen Gliedern, wie z.B. von den Eckzähnen des Sus Babirussa, nicht fragen, wozu dienen sie? sondern, woher entspringen sie? Man wird nicht behaupten, einem Stier seien die Hörner gegeben daß er stoße, sondern man wird untersuchen, wie er Hörner haben könne um zu stoßen.] (FA, I, 24: 234)

As seen in chapter 2, he rejects the teleological argument that animals are born with certain features for certain predetermined purposes and postulates instead a more creative and interactive role for organisms. Animal forms may arise out of a reaction to given circumstances or from an animal's own desire (FA, I, 24: 235). Animal forms and animal ways of life are by no means static. If the circumstances change, so too may the organism. Indeed, in the foregoing passage, Goethe seems to be criticizing the very notion of an animal being "given" any particular feature.

Even more revolutionary aspects of the principle of compensation arise when Goethe gives examples of animal will and describes the changes in form arising because of it. He departs from many of his contemporaries as well as from later Darwinians in his focus upon the inward drive or will of organisms. Organisms are neither solely restricted to change their forms due to a change in their environment (Lamarck), nor do they vary by chance or spontaneously (Darwin). In an 1822 review essay ("Die Faultiere und die Dickhäutigen, abgebildet, beschrieben und verglichen von d'Alton") on two works by the anatomist d'Alton, Goethe discusses how different animal forms arise. D'Alton compares the fossil of the extinct giant sloth, known as Cuvier's *Megatherium* (figure 2), with three-toed (figure 3) and two-toed sloths (figure 4). And although Wells

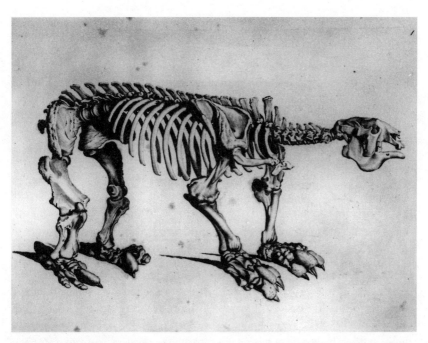

Figure 2: D'Alton's drawing of the extinct, giant sloth, *Megatherium*. Photograph by Angelika Kittel. Reprinted by permission of Stiftung Weimarer Klassik/Goethe-National Museum; Aus Goethes Bibliothek Ruppert No. 4322.

argues that Goethe wanted to give "the principle of transmutation less scope" than d'Alton, and that he praised d'Alton's essay primarily because it "contains a number of studiously favorable references to [Goethe's] own work" (34), a close examination of Goethe's text against d'Alton's shows that Goethe went even further than d'Alton in postulating changes in animal forms. By pointing to the similarities among these animals, d'Alton wishes to argue that nature creates all of its creatures in an uninterrupted chain of being. Goethe, in his review, returns to his idea of the general type (Überzeugung von einem allgemeinen Typus)[19] and gives a different account. He praises d'Alton's juxtaposition of three sloth skeletons, thereby bringing to our attention the "eternal mobility of all forms" [die ewige Mobilität aller Formen] (FA, 1, 24: 546). However, where d'Alton argues that differences in animal forms are due only to the environment (e.g., "Riesen-Faulthier," Einleitung, n.p.), Goethe focuses upon the inner drive of the organism and its *Geist*. Not only is this inner drive so tenacious (hartnäckig) that it maintains the individual character of the organism in the face of enormous change, but it also has the power to change its own form. Goethe had tried to demonstrate with his early botanical treatises that even particular parts of plants exhibit a kind of striving. He similarly attributed such striving to the animal kingdom. He gives us several examples of how organisms, through an exertion of will, may change their own forms and ways of life. Of the examples he gives us, one is a failed attempt, while another a successful one.

Goethe prepares us for his topic first by referring to two pagan gods noted for their ability to change their forms at will: Proteus, the Greek god, and Camarupa, the Indian god of cloud formations (FA, 1, 24: 545). Then he asks for our indulgence. He departs from scientific prose and tells instead a story about an imaginary creature:

One may permit us some poetical expression since prose would in no way be adequate. A monstrous spirit, as one which in the ocean could well show itself as a whale, throws itself upon a swampy, gravelly shore of a torrid zone. It loses the advantages of a fish. It lacks an element to bear its weight—an element which grants the most heavy body ease of movement through the smallest of organs. (My translation)

[Man erlaube uns einigen poetischen Ausdruck, da überhaupt Prose wohl nicht hinreichen möchte. Ein ungeheuerer Geist, wie er im Ozean sich wohl als Walfisch dartun konnte, stürzt sich in ein sumpfig-kiesiges Ufer einer heißen Zone; er verliert die Vorteile des Fisches, ihm fehlt ein tragendes Element, das dem schwersten Körper leichte Beweglichkeit, durch die mindesten Organe, verleiht.] (FA, 1, 24: 547)

Wells argues that Goethe uses the term *poetic* to distance himself from the theory, which is a "fanciful suggestion" (37). An examination of the entire account, especially taken within the context of Goethe's other works on comparative anatomy, shows Goethe's seriousness in supporting the thesis. Goethe's fable resembles an evolutionary account. He describes how life emerged from the ocean and evolved into a new creature.[20] His imaginary creature, however, does not evolve due to a change in the environment. Rather, it chooses a new environment by throwing itself up onto the shore. Nor does this choice stem from a process of evolutionary progression—of a survival of the fittest. Indeed, at first, this creature is more disadvantaged than it was in its life in the water. It cannot move around as readily as before. It therefore acts to compensate for the disadvantages that have resulted from its given choice and grows monstrous or enormous limbs (ungeheuere Hülfsglieder) to help it carry its monstrous body around on land. In the plant, a bicolored petal already symbolized for Goethe nature's drive to overcome current form. Within the animal kingdom, he speculates that a similar drive could explain evolution of form. Because this new creature has lost its previous advantage of a fish in water without as yet having the advantages of a land creature, Goethe characterizes its existence as a kind of slavery (Sklaverei) in which the creature is caught between two environments and is hampered by both. Of course, Goethe's account is completely fanciful in that the changes that he describes are too large to take place within one generation. Like a fairy tale, the time frame is not openly addressed. Nevertheless, within the account, he emphasizes that the impetus to change comes from within the animal and is not due to a change in the environment.[21] Where a mechanist today searches for the causes of the change among genes, Goethe, the vitalist, attributed the change to a type of willing that he saw in various forms throughout nature.

Nor does this particular animal progress in a positive manner, but its own will shapes an imbalance of form that creates even greater impediments for itself. Goethe then describes how, in subsequent generations, this creature transforms itself into an Ai or a three-toed sloth (see figure 3). The Faustian desire of the Ai to overcome its new obstacles on land is so uncontrolled that its long nails signal a desire to continue to grow and expand itself—a desire that seems to have no limit (scheint keine Grenze zu haben). This creature, however, has not budgeted its funds wisely. Its striving is disproportionate with its material qualities. Its disproportionately long arms and legs develop from "impatience" (FA, 1, 24: 547). Similarly, it multiplies and produces its own vertebrae in an attempt to extend its bodily proportions. The cost of its impatience is severe. Not only do its gangly limbs inhibit its freedom of movement so that it is even more restricted than before, but its numerous vertebrae have left little material for its head, which is consequently almost brainless. This unfortunate creature represents a type of monstrous retrogression and shows us by a negative example how the budget works. Because it has attempted to overcome too much too quickly, it not only fails in overcoming the obstacles presented to it, but it creates new, almost insurmountable ones as well. Goethe's description of the evolution of this

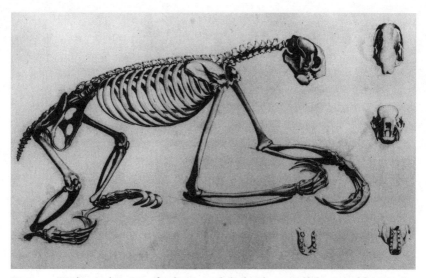

Figure 3: D'Alton's drawing of a three-toed sloth. Photograph by Angelika Kittel.
Reprinted by permission of Stiftung Weimarer Klassik/Goethe-National Museum; Aus
Goethes Bibliothek Ruppert No. 4322.

creature also separates him from Lamarck's view of evolution, which involved "a steady, largely linear advance from 'primitive' to more complex" (Mayr, 430). The individual striving of an organism is not part of a
chain, but indeed breaks through a progressive pattern in order to form a
new entity that is inferior to its earlier manifestation. And although
Goethe postulates an additional, more progressive form, it does not affect the existence of this retrogressive one.

Goethe describes how a second, less hubristic character succeeds
where the first one failed. He continues his tale and describes a second,
more controlled and limited attempt of a creature that tries to live on
land. A type of two-toed sloth, an Unau, achieves a more balanced approach to newfound limitations (see figure 4). Unlike the Ai, the Unau
controls its impatience and, as a result, does not destroy all proportion,
but instead wins physical and intellectual liberation for itself:

It is noteworthy, on the other hand, how in the Unau the animal spirit has controlled itself more, has dedicated itself closer to the earth and accommodated itself to it and has developed towards the mobile ape-genus; as indeed one finds
among the apes several, which could point to it. (My translation)

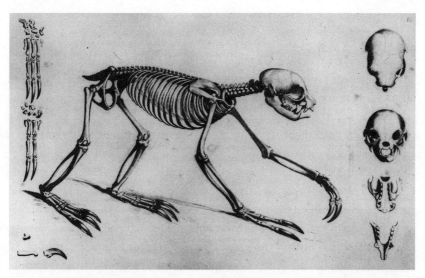

Figure 4: D'Alton's drawing of a two-toed sloth. Photograph by Angelika Kittel. Reprinted by permission of Stiftung Weimarer Klassik/Goethe-National Museum; Aus Goethes Bibliothek Ruppert No. 4322.

[Merkwürdig dagegen ist, wie im Unau der animalische Geist sich schon mehr zusammen genommen, sich der Erde näher gewidmet, sich nach ihr bequemt und an das bewegliche Affengeschlecht herangebildet habe; wie man denn unter den Affen gar wohl einige findet, welche nach ihm hinweisen mögen.] (FA, 1, 24: 548)

Because the two-toed sloths find an equilibrium, they create a new form for themselves that enables physical and intellectual advances. Although the Unau is ultimately more successful than the Ai, both creatures exhibit an inward drive to change their forms and their way of life. Both attempt through an inner drive to reshape their own forms; both therefore demonstrate nature's drive toward diversity. These examples illustrate that inward striving is not necessarily a positive progression. Faust's ultimate salvation occurs when he recognizes the possibility of his own limits. Although he does not want a particular moment to stand still, he is able to imagine the circumstances under which he might be finally satisfied. The Unau, in a similar manner, is a creature that recognizes some limits to its striving and therefore achieves a higher existence than its counterpart, the Ai, which attempts to struggle against all obstacles. Goethe's budget principle gives us a means by which to measure the success of the one creature and the failure of the other, as well as the amount of freedom given to each organism within given limits.

A generation before Darwin, Goethe's fairy tale is revolutionary. It discusses how animal life emerged from the water and later developed into the very mobile apes. Where d'Alton is content to describe the similarities among the Megatherium, the Ai, and the Unau, and account for their differences through a change in the environment, Goethe gives a more bold account of changing forms. His "poetic" tale is not limited, like d'Alton's more scientific account, to discussing similarities among closely related animals. Rather, Goethe links whales, sloths, apes, and perhaps even human beings together, not only because of their progressive development, but also in their abilities to strive to overcome limits. By comparing d'Alton's three drawings of the Megatherium (figure 2), the Ai (figure 3), and the Unau (figure 4), one may better trace the progression that Goethe claims in his review. Once we recall that Goethe linked human beings with the apes through the intermaxillary bone (1786), one could speculate that he would also consider human beings to be part of this course of development. Linnaeus had already "jolted the scholarly world" in 1735 when he placed human beings in the same order as monkeys and sloths (Schiebinger, 80), and Goethe's discovery of the intermaxillary bone made this connection more direct in his mind.[22]

Goethe examines the mobility and the evolution of particular species in a more straightforward manner in other scientific works. For example, in a 1824 review of a d'Alton text illustrating and discussing rodents ("Die Skelette der Nagetiere, abgebildet und verglichen von d'Alton"), Goethe once again focuses upon the malleability of outward form. The rodents, like Goethe's type, serve as a particularly good example of how nature is both free and determined. Without ever losing their rodent characteristics, he describes how rodents may emerge in a multitude of forms. It is striking that where Wells reads d'Alton's work on rodents as stepping back from his earlier, more evolutionary claims and stressing instead the "uniqueness of man" (39), Goethe uses this same work to bring up important similarities between human beings and rodents.

Goethe describes the teeth of all animals as a "shackle of nature" [Woran die Natur das Geschöpf eigentlich fesselt, ist sein Gebiß] (FA, 1, 24: 632). His examination of the rodent family illustrates how this seeming limitation or constraint enables creativity on several different levels.

And while he agrees with d'Alton that the environment exerts a definite influence upon the eventual evolution of form, Goethe, unlike d'Alton, also attributes other causes, including the organism's own creative urge, to the end product of its form. Similarly, Goethe's review shows that he completely rejects d'Alton's emphasis on blind animal instinct as the source of animal creativity. Some Goethean organisms are creative and highly conscious of the world around them: they at times choose their own environments, create artistic products that appear deliberate, and in the case of human beings and rodents, raise themselves into upright posture.

In his account of the rodents, Goethe focuses upon their constant gnawing and links this activity to several different kinds of creativity. He regards this gnawing as a convulsive passion (fast krampfhaft leiden-schaftliches [FA, 1, 24: 633]) or as a "continuous exercise, a restless drive to be occupied" [als fortgesetzte Übung, als unruhiger Beschäftigungs-trieb] that finally culminates in a destructive fit (der zuletzt in Zer-störungskrampf ausartet [FA, 1, 24: 635]). Their destructive gnawing, however, also leads to creative acts. They build their living spaces with the gnawed materials, and once they have satisfied their needs, they exert their energy toward the collecting drive (Sammlertrieb). This drive, in turn, leads to actions that Goethe likens to deliberate artistry (überlegte Kunstfertigkeit [FA, 1, 24: 635]).[23] In an essay on the Cuvier-Geoffroy debate, Goethe even more explicitly describes what this deliberate artistry could be—and even more explicitly links it to human characteris-tics: he sees structures built by beavers as analogous to rational architec-ture (vernünftige Architektonik [FA, 1, 24: 835]).

Goethe, however, does not limit rodent artistry to their outward ac-tivities, but argues that they have an influence upon their own bodily forms. While Goethe admits that animal forms are in part determined and static, he also argues that the forms can change and transform them-selves into infinite varieties and even new animals over the generations (FA, 1, 24: 632). As in his essay on sloths, Goethe stresses the versatility of this animal group and its ability to use its own will (Willkür) to effect these changes. To stress the extreme malleability of the rodent form, he further explains how various rodents bear similarities to very different kinds of animals. The rodent order "leans as much toward the carnivores as toward the ruminants, as much toward the apes as toward the bats, and resembles still others species which lie in between" (my translation) [wie es sich denn sowohl gegen die Raubtiere als gegen die Wiederkäuer hinneigt, gegen den Affen wie gegen die Fledermaus, und noch gar andern dazwischen liegenden Geschlechtern sich anähnelt] (FA, 1, 24: 635). Although he does not overtly state that such different animals as apes, bats, and squirrels could have the same ancestor from which they then evolved, he emphasizes the similarities among disparate animal groups. Moreover, Goethe takes the very bold step of linking one of the signature characteristics of human beings—upright posture—to squirrels and other animals as well:[24]

for otherwise one notices overall in the four-footed animals a tendency of the posterior extremities to raise themselves above the frontal ones, and *in this we believe that we glimpse the basis of pure upright posture of human beings.* How such striving can, however, gradually intensify to disproportion is especially evident in the rodent genus. (My translation, emphasis added)

[denn sonst bemerkt man überhaupt an den vierfüßigen Tieren eine Tendenz der hintern Extremitäten sich über die vordern zu erheben, und *wir glauben hierin die Grundlage zum reinen, aufrechten Stande des Menschen zu erblicken.* Wie sich solches Bestreben jedoch nach und nach zur Disproportion steigern könne, ist bei dem Geschlecht der Nager in die Augen fallend.] (FA, 1, 24: 633–34, emphasis added)

Thus, when a squirrel stands up (figure 5), its striving is different from that of human beings only in degree and proportion.[25] Sometimes, the striving results in success: an animal breaks free from some restraints. Such a creature is the human being, who ascended from four-footedness to upright posture. Other times, the striving results in failure. The rodent type develops a disproportionate body and loses freedom of movement. Each of these successes, as well as each of these failures, is symbolic of how the animal has balanced its budget given its particular outward circumstances and inner drive. Successful striving to overcome some limits is directly linked to the proper balancing between the extremes of fixed limits and capriciousness. This striving, moreover, becomes the central focus in Goethe's evolutionary views.

A brief comparison between Goethe's version of upright posture and Lamarck's (1809) further emphasizes the differences between the two men. Lamarck postulates Newtonian-type causes to explain the end effect of upright posture:

Figure 5: D'Alton's drawing of an upright squirrel. Photograph by Angelika Kittel. Reprinted by permission of Stiftung Weimarer Klassik/Goethe-National Museum; Aus Goethes Bibliothek Ruppert No. 4326.

If some race of quadrumanous animals, especially one of the most perfect of them, were to lose by force of circumstances, or some other cause, the habit of climbing trees and grasping the branches with its feet in the same way as with its hands in order to hold them, and if the individuals of this race were forced for a series of generations to use their feet only for walking and to give up using their hands like feet there is no doubt . . . that these quadrumanous animals would at length be transformed into bimanous, and that their thumbs on their feet would cease to be separated from the other digits when they use their feet for walking and that they would assume an upright posture in order to command a large and distant view. (Quoted in Mayr, 352)

Where Goethe emphasizes the internal striving of an animal as the impetus to change, Lamarck postulates that such a change occurred through external causes.[26] For Lamarck, a change in the environment or in external circumstances had to occur before there was a change in the organism. Moreover, these changes would then occur in a species across the board and would not originate within an individual. Lamarck postulates a series of physiological changes, based upon Cartesian mechanisms, to explain the physical changes (Mayr, 354). Thus, where many of

Goethe's contemporaries believed in a chain of being, where species as a group gradually changed and become more perfected, Goethe locates that change within the individual and even further rejects an ever-perfecting chain by postulating new, but retrogressive, organisms, such as the Ai. Goethe's version of organic evolution is also certainly different from Darwin's in that Darwin's notion of natural selection is quite Newtonian: he turns to a mechanism, natural selection, to explain evolution. Darwin, moreover, relegates to the back burner the question of the causes behind variations that bring natural selection into play (Behe, 6; Mayr, 682–83).

Goethe and Evolution

In the second half of the nineteenth century, Ernst Haeckel (66, 75–76, 79) argued that Goethe's morphological theories were precursors to Darwinian ones. Ever since Haeckel's statements, a controversy has raged over whether and to what extent Goethe prefigured Darwin. Did Goethe envision a descent of the species? Did he believe in common ancestry?[27] Several twentieth-century scholars have answered these questions in the negative.[28] Some scholars, such as D. Kuhn (15; FA, 1, 24: 991–92), Wells (1), Wenzel ("Goethes Morphologie," 62–67), and, of course, Foucault (150–53)[29] argue, however, that the notion of evolution was completely foreign to someone in Goethe's time and that it was highly unlikely that anyone in Goethe's historical era could have conceived of such a concept. Because scholars such as D. Kuhn, Wells, and Wenzel focus upon the more narrow aspects of Goethe's evolutionary thinking (that he only seemed to apply it to vertebrates), they seem to miss the great fluidity of Goethe's theory as well as the active role that organisms play in their own development (as Lenoir and Cornell have already noted).[30]

Goethe's evolutionary thinking, evidenced by his comparative anatomy and morphological essays, goes well beyond reactive or spontaneous change. Organic change is part of a large and complicated relationship that includes an animal's will, its given budget, and environmental influences. For Goethe, survival of the fittest would therefore be an inadequate measure of an animal's existence and natural selection a too mechanical explanation that does not really explain the root cause of change. Goethe was certainly not a precursor to Darwin in that Goethe approached evolution from a different perspective. Darwin's theory aimed at ridding evolution of all nonmaterial causes. Goethe had already rejected the Newtonian causal explanations of his contemporaries. Such theories in Goethe's mind did not take the individual striving of an organism into account and represented too narrow a view of how nature develops. Goethe's view of nature was every bit as competitive as Darwin's. Nature in Goethe's scientific texts at times may be brutally competitive (chapter 4). However, Goethe argues that more than survival was at stake for organisms in their struggles. His natural philosophy includes a striving toward flourishing and beauty outside of issues of survival. For example, his essay "Fossil Steer" ("Fossiler Stier," FA, 1, 24: 554–60), which critics admit comes the closest to advocating a descent of the species (FA, 1, 24: 1108), discusses a fossil of an extinct steer and compares its form to its more modern descendants.[31] Goethe's main interest in the discovery of this fossil, however, is not in its modern descendants, although he clearly has an interest in the development of its morphological form. His focus is on the beauty of its horns (FA, 1, 24: 558). He discusses how their curved form is but one aspect of living nature. Nature, whenever it has some superfluous amount of material or energy, has a tendency to curve and curl. Instead of linking this beautiful feature to a factor in its survival, Goethe notes that the very thing that makes the horns beautiful makes them less useful as weapons (FA, 1, 24: 559). In addition, when discussing the shape of its horns, he points to how outside forces working together with internal ones may shape them. For example, human beings can change the shape of the horns by applying a brace to them as they grow. Nature's will to curl and curve is enhanced by an external impediment (FA, 1, 24: 559–60). Thus, Goethe here emphasizes the impact of environmental forces upon form and upon nature's desire to create beauty with its excess resources.

By briefly returning to Goethe's accounts of sloths and squirrels, his intimations of evolutionary views become more apparent. In discussing the transformation of species, he does not limit his observations to closely related groups. The whale becomes a sloth, which then turns ape-like. Notably in Goethe's "fable," life begins in the sea only to evolve into more complex forms on land.[32] At each stage of the whalelike creature's development, its will plays a central role. It beaches itself, it strives to overcome obstacles, it suffers from impatience, and it ultimately through balanced striving finds a better form. Goethe similarly portrays the rodent species as extraordinarily fluid. It is able to sway in its developmental forms to resemble carnivores, ruminants, apes, and bats.

That Goethe even links upright posture, a supposedly distinguishing human feature, to squirrels demonstrates how radical his thinking could be and how seriously he took the concept of the unity of nature. In discussing a squirrel's manner of eating, he further attributes table manners and forethought to it. He tells us that the squirrel is the highest formation of its kind (höchste Ausbildung) and has a particular skillfulness in delicately (zierlich) handling small, appetizing (appetitliche) objects. It mischievously (mutwillig) seems to play with its nut (Nuß) as a means of preparing for its enjoyment (Genuß) of it (FA, 1, 24: 634). Goethe's wordplay shows how carefully he constructed this passage to demonstrate the unity of nature. We are not projecting our own selves upon the squirrel when we see it acting in ways resembling human activities. The general desires, although different in degree, really are the same throughout nature.

Of course, this question arises: If Goethe indeed had such original and evolutionary views, why did he not more openly discuss them? He himself gives us several indications within his morphological writings that he was intentionally circumspect on this issue. As Wenzel as already noted, Goethe uses ambiguous terms to discuss his type in precisely those passages where he discusses the scope of its application ("Goethes Morphologie," 60). Nor is this ambiguity limited to his essays on comparative theory. By examining those texts in which Goethe comes the closest to endorsing modern evolutionary views, one discovers several instances of strangely veiled or ambiguous language.[33]

In the essay on rodents, Goethe touches upon some of the difficulties of openly advocating an evolutionary theory. In discussing the ability of the rodents to transform themselves and sway from form to "unform," he raises the possibility that honest observers (redliche Beschauer) will be placed into a kind of insanity (eine Art von Wahnsinn [FA, 1, 24: 633]). This insanity may well refer to the reaction one may have in thinking about evolution for the first time. Indeed, the following line may be read as a reference to Galileo's troubles with the church: "because for us limited beings it might well be better to remain fixed in error than to sway with the truth" (my translation) [denn für uns beschränkte

Geschöpfe möchte es fast besser sein den Irrtum zu fixieren als im Wahren zu schwanken] (FA, 1, 24: 633). As Galileo is said to have muttered under his breath, "But it moves," so too does Goethe's entire essay admonish us to be as malleable as nature in our study of it, as it changes and evolves from one form into another, and not to be fixed. Of course, Goethe was not by any means under the same pressures as Galileo when it came to censorship. However, it is also interesting to note that even Darwin waited twenty years before publishing his work because of his fear of the reaction that it would cause (Mayr, 420). Goethe does, however, give us an indication in another essay as to why he might have been careful to publicize his evolutionary theories.

Goethe himself uses the word dangerous when discussing some of the paths that lead from his theory. He uses the term not to refer to his own possible problems, but to the impact that evolutionary theory might have on a general reading public. In "Problems" (1823), he uses the term metamorphosis broadly to refer to the changes in organic formation.[34] He calls the "idea of metamorphosis an extremely honorable one" but also an "extremely dangerous gift from above" [Die Idee der Metamorphose ist eine höchst ehrwürdige, aber zugleich höchst gefährliche Gabe von oben] (FA, 1, 24: 582). The idea is dangerous because it "leads to formlessness; it destroys knowledge" [Sie führt ins Formlose; zerstört das Wissen] (FA, 1, 24: 582). It acts like a centrifugal force and would even lose itself in the infinite had it not an opposing force; the "drive for specific character, stubborn persistence of things which have finally attained reality" (12: 43) [den Spezifikationstrieb, das zähe Beharrlichkeitsvermögen dessen was einmal zur Wirklichkeit gekommen] (FA, 1, 24: 582–83). This formlessness is akin to a loss of belief in a type of structure or overall organization. He argues that when some beliefs are challenged or overturned, human beings create new ones to replace them [Wir müßten einen künstlichen Vortrag eintreten lassen. Eine Symbolik wäre aufzustellen] (FA, 1, 24: 583). This symbolism, however, has far-reaching consequences. Human beings create their symbols and their own language for things, whether for religion, government, or art, in order to escape from order-destroying ideas. In other words, Goethe views certain scientific methods as so destructive to the human sense of order that human beings need to create artifices to suppress nihilistic knowledge.

Goethe perhaps most fully examines the issues of evolutionary thinking in his poem "Metamorphosis of Animals." The poem's discussion of
the place of human beings within nature suggests an evolutionary view.
It begins with daring its readers to ascend to the last step of nature:

> Now if your mind is prepared to venture upon the final
> Step to the summit, give me your hand and view with an open
> Gaze the abundance of Nature before you.
> (1: 161)

> [Wagt ihr, also bereitet, die letzte Stufe zu steigen
> Dieses Gipfels, so reicht mir die Hand und öffnet den freien
> Blick ins weite Feld der Natur.]
> (FA, 1, 2: 498)

The first lines imply that the observer of nature must have some kind of
courage or be willing to take some kind of risk in his or her study of nature's changing forms. This challenge involves the willingness to see new
vistas and one's place in nature from a new perspective. Nature is not a
caring mother who looks out for her children, but simply provides the
structure for life and creative production through the principle of compensation. The natural observer, therefore, needs primarily to be aware
of this principle in order to study the metamorphosis of form. The new
perspective that is opened up to the observer, as the concluding lines of
the poem demonstrate, requires the observer to look backward, investigate, and compare:

> Stand where you are, be still, and looking behind you, backward,
> All things consider, compare and take from the lips of the Muse then,
> So that you'll see, not dream it, a truth that is sweet and certain.
> (1: 163)

> [. . . . Hier stehe nun still und wende die Blicke
> Rückwärts, prüfe, vergleiche, und nimm vom Munde der Muse
> Daß du schauest, nicht schwärmst, die liebliche volle Gewißheit.]
> (FA, 1, 2: 500).

Readers are to see themselves—although for the moment as a fixed
form—not only as a product of generations of change, but also as a step-

ping stone toward future development. Human beings, like nature as a whole, become both fixed and free. Our forms, like those of the animals at large, are products of a long chain of development.

The scientific method that Goethe here advocates also mimics nature's activities. The natural observer stands still, but then turns his or her gaze around. He or she must be both static and mobile. In the poem, the principle of compensation is just as applicable to the development of morphological forms as is it to politics, art, and science. The principle of compensation represents a delicate balance (similar to the balance between environmental influences, animal parts, and will) between opposing forces:

> May this beautiful concept of power and limit, of random
> Venture and law, freedom and measure, of order in motion,
> Defect and benefit, bring you high pleasure; gently instructive,
> Thus, the sacred Muse in her teaching tells you of harmonies.
> Moral philosophers never attained to a concept sublimer,
> Nor did men of affairs, nor artists imagining; rulers,
> Worthy of power, enjoy their crowns on this account only.
> (1: 163)

> [Dieser schöne Begriff von Macht und Schranken, von Willkür
> Und Gesetz, von Freiheit und Maß, von beweglicher Ordnung,
> Vorzug und Mangel, erfreue dich hoch; die heilige Muse
> Bringt harmonisch ihn dir, mit sanftem Zwange belehrend.
> Keinen höhern Begriff erringt der sittlich Denker,
> Keinen der tätige Mann, der dichtende Künstler; der Herrscher,
> Der verdient es zu sein, erfreut nur durch ihn sich der Krone.]
> (FA, I, 2: 193)

The unity of nature therefore extends beyond linking the human form with those of the animals. Goethe extends the principle of compensation into the realm of the most human of activities. Human beings are to take their cue from nature as they engage in politics, aesthetics, and morality. By emphasizing that we are a creature of nature even in our most human of endeavors, he calls our attention to the fact that our own physical development must also mirror that of the other creatures. Our position as the highest creature (illustrated by our position on top of a mountain) gives us the perspective to see the true working of nature as well as our place within it. By looking backward at nature, one sees perhaps not only natural history, but our own history.

Inanimate Nature and the "Will" to Create

Until now, this chapter has been investigating the principle of compensation in the animate world. Goethe, however, also endowed inanimate natural phenomena with a creative drive. In *Elective Affinities,* one of his characters, the Captain, speculates that matter itself is alive with desire, so that its interactions become tales of love, friendship, and hatred. The Captain, in explaining the crosswise pattern of some chemical reactions, observes,

In this kind of separation and combination, repulsion and attraction, some higher destiny really does seem to manifest itself; one almost feels like endowing such substances with the ability to exercise choice and will, so that the term 'elective affinities' seems perfectly justified. (II: 116)

[In diesem Fahrenlassen und Ergreifen, in diesem Fliehen und Suchen, glaubt man wirklich eine höhere Bestimmung zu sehen; man traut solchen Wesen eine Art von Wollen und Wählen zu, und hält das Kunstwort 'Wahlverwandtschaften' für vollkommen gerechtfertigt.] (FA, 1, 8: 305)

The Captain speaks of the elements as if they have their own wills because they appear to act as if they do. Goethe echoes these sentiments in his scientific works when discussing several different inanimate entities. For example, in chapter 2, I argued that for Goethe pure red illustrates how colors strive for a higher level of complexity. Even more striking, however, is how he describes clouds in his meteorological writings. Here, he speaks over and over again of the phenomena of meteorology as something living (etwas Lebendiges), that one daily sees acting and creating (FA, 2, 12: 309). As noted earlier, Goethe viewed himself as a Lucretian—as a natural philosopher who saw striving even within nature's inanimate parts. However, where Lucretius endows every atom with the ability spontaneously to swerve at any given moment, Goethe endows natural entities, whether animate or inanimate, with the ability to flourish and strive for more particular, albeit changing, ends. Where the Cartesian world of dead matter operates according to set rules, Goethean matter comes to life. Goethe's meteorological works, belonging to his late period, further illustrate the continuity of his thoughts on the principle of compensation. Within these works, one can further follow the relationship between polarity and Steigerung that leads to the creation of more perfect forms. Inanimate nature, like its animate counterpart, is best able to be creative when limited. In addition, the meteorological works lay out more specifically Goethe's notion of the ideal scientific method, which is based primarily upon nature's dynamic patterns.

Goethe's meteorological writings have two recurring themes: (1) the ever-changing nature of clouds, and (2) praise for Luke Howard, the man who invented the cloud nomenclature still in use today (cirrus, cirrocumulus, cumulus, stratocumulus, nimbus, stratus, etc.).[35] These themes are very closely related. Goethe's warm praise for Howard's method stems from Howard's recognition of nature's creative powers and his ability to create a scientific system that reflects its dynamism. Howard's system, like Goethe's own type, recognizes nature's inherent malleability and creates a methodology around it.

In "Cloud Shapes According to Howard" ("Wolkengestalt nach Howard," 1820), Goethe characterizes his own study of particular cloud formations as a wonderful, highly instructive play (ein herrliches, höchst unterrichtendes Schauspiel [FA, 1, 25: 220]). As in the introduction to The Theory of Colors, nature creates a rich theatrical production. This time, however, clouds rather than colors play the leading roles. Like organic entities, they are unpredictable, constantly changing, and superficially appear to have no limits. A close investigation of them, as of any natural phenomena, reveals that their creative forms follow certain dynamic patterns and arise out of the limits presented to them under various circumstances.

Clouds are active and changeable in part because they are part of a large polar battle (Konflikt) between the earth and the sky (FA, 1, 25: 216). The earth exerts a material pull upon the clouds and, accordingly, the closer that they are to the earth, the more material and water-laden they become. Conversely, as the sky pulls the clouds upward, it causes their more material qualities to dissipate until the cloud itself disappears into the atmosphere. Clouds represent a middle ground between the two primal forces of earth (or the material pull) and the sky (or the refinement of the more spiritual forces). At each end of the spectrum, one sees the destruction of form. Too much of the material influences causes the cloud to rain down and lose its form, whereas too much of the heavenly influences causes the whole form to become so spiritual as to disappear.

While the cloud seems simply to be tossed about by atmospheric forces, these actors on the stage of the sky not only suffer (e.g., FA, 1, 25: 200) but also act.[36] They are not only passive participants in the conflict between upper and lower regions, but also actively engage in this conflict (FA, 1, 25: 220). Although clouds are faced with limits to their being, they still manage to have a certain amount of freedom and creativity. They are always creating new shapes within their particular cloud formations (cumulus, nimbus, etc.), they seem to determine some of their own movements apart from being pulled by atmospheric influences, and they can maintain some of their own character as they transform from one cloud form to the next.

Goethe endows cumulus clouds with a special status.[37] They occupy the middle region between earth and sky and therefore are in the middle of the great conflict between these two forces. This middle position, however, also opens up the possibility for an ideal reconciliation of these opposing pulls. Their very shape manifests this special position. They contain some of the characteristics of the clouds beneath them and of the clouds above (FA, 1, 25: 231–32). According to Goethe, they are the most beautiful of clouds and the most deserving of the very name *cloud* (FA, 1, 25: 200).[38] And while he speaks of all clouds as having a certain agency, cumulus clouds seem better able to maintain their own characters throughout their various changes and exhibit the most independence of formation. Just as yellow and blue at times seem to intensify from within to form pure red, so too do cumulus clouds act from within to create their outward form. They raise themselves up, follow their own movements, and make their own ball-like forms (FA, 1, 25: 200, 208). For example, in a study of the weather on 28 April 1820, Goethe notes the changes of the cumulus clouds as they shift from one form into another. While it is at one point forced to rain, part of it still maintains its identity, rises up again, joins with various stratuslike clouds, and then returns to its original state (FA, 1, 25: 220–21). This cloud, although undergoing several transmutations, is still able to maintain part of itself—its shape or Gestalt—throughout the change.

Goethe repeatedly compares clouds to Camarupa, the Indian god. He became familiar with this mythic figure after reading H. H. Wilson's 1813

English translation of Kalidasa's epic poem, *Mega Dūta* or *The Cloud Messenger*.[39] Goethe characterizes this god as one who delights in the changing of forms (FA, 1, 25: 199) and who is the "wearer of shapes at will" (FA, 1, 25: 243).[40] These changing forms explain why artists and poets are so fascinated by clouds (FA, 1, 25: 214, 242): it is a "function of the human imagination" and a "natural instinct" (FA, 1, 25: 243) to want to fix their forms by drawing them on a page or by comparing them to more concrete objects (animals, armies, etc.). The human mind is often confused by this myriad of forms and change (FA, 1, 25: 237), or it lets its imagination run wild (FA, 1, 25: 242, 243). Goethe distinguishes the poetic treatment of nature from a scientific one. Where a poet uses his or her imagination to "give a necessary form to every shapeless image of chance" [allem ungebildeten Zufälligen jederzeit irgend eine notwendige Bildung zu geben trachtet] (FA, 1, 25: 242, 243), the scientist must in some way provide a way to delineate and categorize these forms without losing sight of their fluidity. The manifold forms of clouds make it difficult to distinguish one form from another or to map the cyclical paths of the clouds as they change from one into another. Scientists studying meteorology, therefore, have the same problems as those who wish to investigate comparative anatomy without employing the type: they have no fixed point of reference.

Goethe's cloud essays and poetry focus on the difficulties that one has in analyzing these elusive beings. He notes that his very desire in the case of the clouds to see "the formation of the formless—a lawlike shape-change of the unlimited" [die Formung des Formlosen, ein gesetzlicher Gestalten-Wechsel des Unbegrenzten] (FA, 1, 25: 235) is part of all of his scientific and artistic endeavors. The study of clouds highlights one of the main difficulties that scientists have in studying nature as a whole. How does one begin to analyze something that is always in flux and motion? An examination of his specific praises for Luke Howard makes even more clear how important it is for Goethe to keep nature's dynamism in mind while engaging in scientific research. Howard's terminology functions like Goethe's morphological type in that it gives the natural observer a fixed and more limited point of departure, while still focusing upon nature's malleability.

To Goethe, Howard's success lay in his ability to distinguish the clouds (FA, 1, 25: 237) and categorize "the manifold forms of clouds through terminology" [die mannigfaltigen Formen der Wolken durch Bennenung sonderte] (FA, 1, 25: 199). All nations, according to Goethe, ought to adopt Howard's terms so that scientific discussions would be made easier (FA, 1, 25: 233). Howard's method enabled Goethe to detect the interrelationship of various atmospheric elements (clouds, barometric pressure, temperature, etc.) and to mark their reciprocal influence upon one another. Howard's terms provide somewhat artificial and fixed points from which to compare these changing forms and thus provide guideposts for comparisons (FA, 1, 25: 235). These terms then enable one momentarily to fix a changing nature—to hold it fast: "That which no hand can reach, no hand can clasp, / He first has gain'd, first held with mental grasp"[41] [Was sich nicht halten, nicht erreichen läßt, / Er faßt es an, er hält zuerst es fest] (FA, 1, 25: 238). Howard's method enables scientists to exist in a Faustian moment because they are able to hold fast to a given moment.[42]

Goethe elevates Howard's accomplishments to a philosophical level and praises his personality. As noted earlier, he seems to overemphasize the importance of Howard's life. Goethe claims that only a particular type of person could have accomplished what Howard had done. For Goethe, Howard is the type of man to whom nature shows itself without reserve and with whom it is inclined continuously to converse (FA, 1, 25: 236). He is further the man who "Defin'd the doubtful, fix'd its limit-line, / And nam'd it fitly [Bestimmt das Unbestimmte, schränkt es ein, / Benennt es treffend] (FA, 1, 25: 238, 239). Goethe was so interested in Howard's method that he commissioned a friend in London to gather biographical data on Howard, as "everything that occurs through human agency must be observed in an ethical sense" [daß alles was durch den Menschen geschieht in ethischem Sinne betrachtet werden müsse] (FA, 1, 25: 235). As Goethe did not believe in purely objective science, he was interested in knowing what aspects of Howard's life had given him his particular natural perspective. In particular, Goethe was interested in what part of Howard's education or development enabled his mind (Geist) to "observe nature naturally, give itself over to it, recognize its laws, and to prescribe them back to nature in a natural-human way" [die Natur natürlich anzuschauen, sich ihr zu ergeben, ihre Gesetze zu erkennen und ihr solche naturmenschlich wieder vorzuschreiben] (FA, 1, 25: 236). Goethe believed to have found a kindred spirit in Howard.

Just as Goethe attempted in his own science to reflect as much of nature's processes within his methodology, so too did he believe to recognize a similar methodology within Howard's work. And, just as Goethe published autobiographical information alongside his own scientific works in order to make his perspective clear to others, so too did he also publish Howard's autobiographical statement for the benefit of all "lovers of the truth" (Wahrheitsliebenden [FA, 1, 25: 235]).

Goethe's joy in Howard's nomenclature illustrates Goethe's own understanding of nature and his advocacy for a more dynamic approach to science. If we go beyond Goethe's panegyric praise of Howard, we see that Howard's categories for clouds pass Goethe's litmus test for science. They represent and portray nature as it appears, and they focus our attention upon visible phenomena rather than abstract formulas. By using the term *naturmenschlich* (*Natur* = nature and *menschlich* = human) to describe Howard's method, Goethe stresses that, although Howard's terminology is certainly a human construct, it is not entirely divorced from nature, but is based upon it. In addition, Howard's method encourages one to see nature as a constantly repeating, yet dynamic, cycle. Like Goethe's own "Metamorphosis of Plants" or "Metamorphosis of Animals," the cyclical aspect of the object of study encourages scientists to move forward and backward in their investigations. Finally, Howard's terms enable the observer to study nature's transitional forms. Throughout Goethe's scientific works, he stresses transitional forms as a means of uncovering nature's dynamic patterns. Just as the bicolored tulip leaf or petal enabled Goethe to see the transitional change of the organ of the leaf, so too does Howard's method call attention to nature's cycles and transitions.

Goethe's own meteorological research demonstrates how he incorporated the idea of limiting or fixing nature within his own study of it. While Goethe's essays on comparative anatomy give examples of holding nature fast with a type, Goethe's meteorological essays give examples of what it might mean to "fix" or "grasp" inanimate nature. In his essay "Toward a Theory of Weather" ("Versuch einer Witterungslehre," 1825), Goethe often uses terms that describe a desire to grasp or to hold nature fast. For instance, he writes about our desire to grasp (begreifen) the truth, even though we apprehend it as incapable of being grasped (unbegreiflich). He explains that this desire applies to gaining knowledge of all phenomena of the tangible world (der faßlichen Welt), although in this particular case he will speak only of the difficult to touch/grasp (schwer zu fassenden) area of meteorology.

He then describes some of the general difficulties that scientists have in studying a nature that is composed of intricate and complex relationships. Scientists must always be aware of these complexities:

Above all we must remember that nothing exists or comes into being, lasts or passes, can be thought of as entirely isolated, entirely unadulterated. One thing is always permeated, accompanied, covered, or enveloped by another; it produces effects and endures them. And when so many things work through one another, where are we to find the insight to discover what governs and what serves, what leads the way and what follows? This creates great difficulty in any theoretical statement; here lies the danger of confusion between cause and effect, illness and symptom, deed and character. (12: 145–46)

[Hier ist nun vor allen Dingen der Hauptpunkt zu beachten: daß alles was ist oder erscheint, dauert oder vorübergeht, nicht ganz isoliert, nicht ganz nackt gedacht werden dürfe; eines wird immer noch von einem anderen durchgedrungen, begleitet, umkleidet, umhüllt; es verursacht und erleidet Einwirkungen, und wenn so viele Wesen durch einander arbeiten, wo soll am Ende die Einsicht, die Entscheidung herkommen was das Herrschende was das Dienende sei, was voranzugehen bestimmt, was zu folgen genötigt ist? Dieses ists, was die große Schwierigkeit alles theoretischen Behauptens mit sich führt, hier liegt die Gefahr Ursache und Wirkung, Krankheit und Symptom, Tat und Charakter zu verwechseln.] (FA, 1, 25: 275)

Here Goethe lays out why Newtonian or Cartesian approaches are inadequate means by which to study nature. It is a mistake to approach nature as if it were a straightforward ratio of related parts or a linear equation where one term follows the other. Rather, the scientist must recognize not only that all natural parts have a relationship to one another, but that their very relationship is a complicated one, where it is even difficult to distinguish the cause and the effect.

Faced with the problem of an ever-changing, intricately interconnected object of study, Goethe suggests the following scientific methodology:

The serious observer has no choice but to choose some midpoint and then see how he can deal with what is left on the periphery. This is what we have attempted to do, as the following will show. (12: 146)

[Hier aber bleibt für den ernst Betrachtenden nichts übrig, als daß er sich entschließe irgendwo den Mittelpunkt hinzusetzen und alsdann zu sehen und zu suchen, wie er das Übrige peripherisch behandle. Ein solches haben auch wir gewagt wie sich aus dem Folgenden weiter zeigen wird.] (FA, 1, 25: 275)

The method which Goethe advocates is not that different in theory from the concept of his own type, which he employs to study animate nature. The language is also reminiscent of the scientist's stance in "Metamorphosis of Animals." The scientist should be both mobile and fixed. He or she must arbitrarily pick a point of departure as a static reference point. Then, however, the scientist must be willing to move from that point in order to investigate a nature that is itself in flux. Nor is this point fixed in time, but scientists must be willing to move backward and forward in time (FA, 1, 25: 298–99) in their investigation. While scientists have to be willing to fix a point to study a changing nature, Goethe warns that they cannot forget either to move all around that point or that that point itself ought to be periodically moved in order to gain new perspectives on nature. Because scientists directly determine the particular perspective by choosing their point of departure, they must also not forget their own role within their research. Natural observers must test themselves in a variety of ways (FA, 1, 25: 299) to maintain close contact with nature.

In this essay, Goethe stresses the necessity of placing limitations upon oneself. Our natural relationship to nature is to try to limit, circumscribe, and control it. He explains that human beings have always struggled against natural forces and that the only way to subdue something that is unruly is to use the force of law: "Nature has prepared the way for us in the most wonderful fashion by setting an alive, formed existence against the formless" (12: 147) [hier hat uns die Natur aufs herrlichste vorgearbeitet und zwar indem sie ein gestaltetes Leben dem Gestaltlosen entgegen setzt] (FA, 1, 25: 295). He implies that human beings constantly strive to subdue wild, natural forces (one is here reminded of Faust's land reclamation project) with the power of our mind (Geist), through courage (Mut) and cunning (List). Human beings, however, must learn from nature how to subdue it. The weather gives us numerous examples of how nature limits itself and strives to find a balance of forces. Although the four main elements (air, earth, wind, and fire) are wild and brutal and, indeed, capriciousness (Willkür) itself (FA, 1, 25: 295), these forces eventually find an equilibrium among themselves through a dynamic principle of balance (FA, 1, 25: 296–97). High barometric pressure restrains the elements, while low barometric pressure lets them loose. After long periods under one condition, nature will eventually find its equilibrium by having one force limit the other (FA, 1, 25: 296). The principle of compensation eventually balances out competing forces because it brings about an equilibrium.

Goethe's meteorological writings reinforce his views on the unity of nature. Where the principle of compensation rules over the creation and evolution of animal forms, it also is one of the main forces behind the actions of inanimate nature. Cloud formations and changes in weather patterns may appear chaotic and fluid, but they are governed by the principle of compensation. One force always gains at the expense of another, and an equilibrium brings about the most ideal forms and conditions.

The Aesthetics of Compensation

An underlying theme throughout this chapter has been nature's ability to balance a variety of forces and materials: animal wills engage in a creative struggle against material and environmental limitations, some clouds struggle to maintain their character as they are pulled in a variety of directions, and Goethe's own type, like Howard's nomenclature and Goethe's selection of a midpoint, attempts to reconcile the opposing concerns of motion (the activity of nature) and rest (the scientist's attempt to hold nature fast). In addition to describing and accounting for morphological or meteorological changes, Goethe's principle of compensation has significant consequences for his aesthetic theory: nature is most beautiful in those instances in which it has successfully balanced opposing forces. I have already noted references to beauty in his general discussions of the principle, including a discussion of the beauty of the principle itself ("Metamorphosis of Animals") or the steer's horns. His discussions of art and beauty make more evident the aesthetic and philosophical significance of the principle of compensation. He applies the same principle to evaluate his own artistic processes as he does to discuss beauty in the natural world.

The close link between creativity, beauty, and the principle of compensation emerges in Goethe's statements about beauty in the natural world. When he describes an organism's attempt to flourish and strive toward beauty, he stresses its need to balance freedom, superfluity, and its own will against restraint, scarcity, and the obstacles of a given environment. Superfluity of resources is just as damaging to an organism as scarcity. An organism must be able to channel its resources between these two extremes if it is to be beautiful.

In a conversation with Eckermann, Goethe uses the examples of several different oak trees to discuss how balance brings about beauty in nature. To be beautiful, an oak tree must have struggled with opposition and competition as it has grown. If an oak tree grows

in a moist marshy place, and the earth is too nourishing, it will, with proper space, prematurely shoot forth many branches and twigs on all sides; but it will still want the opposing, retarding influences: it will not show itself gnarled, stubborn, and indented; and, seen from a distance, it will appear a weak tree of the lime species; it will not be beautiful—at least, not as an oak. (18 April 1827)

[Wächst hinwieder die Eiche an feuchten, sumpfigen Orten und ist der Boden zu nahrhaft, so wird sie, bei gehörigem Raum, frühzeitig viele Äste und Zweige nach allen Seiten treiben; es werden jedoch die widerstrebenden retardierenden Einwirkungen fehlen, das Knorrige, Eigensinnige, Zackige wird sich nicht entwickeln, und aus einiger Ferne gesehen, wird der Baum ein schwaches lindenartiges Ansehn gewinnen, und er wird nicht schön sein, wenigstens nicht als Eiche.] (FA, 2, 12: 599)

A tree that has never faced adversity will never reach its highest potential. Goethe similarly spoke of an overwatered plant in his "Metamorphosis" essay. Although such a plant will continuously grow leaves, it will never develop flowers or more specialized organs. So, too, does it become a central fact in the plot of *Elective Affinities* that the character Eduard had never been denied anything. His insistence on getting his own way causes the deaths of his son and lover and thereby destroys any hope for his happiness.

Conversely, an oak tree that must struggle too much will waste too much of its energy on simple survival. If it had to strive simply to reach light and air, "its many years of upward striving have consumed its freshest powers" (18 April 1827) [ihr vieljähriger Trieb nach oben hat ihre frischesten Kräfte hingenommen] (FA, 2, 12: 598), so that it is unable to strive toward beauty. Moreover, if it grows upon poor soil, it will also develop quite differently:

If, lastly, it [the oak] grows upon a mountainous slope, upon poor soil, it will become excessively gnarled and knotty; but it will lack free development; it will become prematurely stunted, and will never attain such perfection that one can say of it, "There is in that oak something that creates astonishment." (18 April 1827)

[Wächst sie [die Eiche] endlich an bergigen Abhängen, auf dürftigem steinigten Erdreich, so wird sie zwar im Übermaß zackig und knorrig erscheinen, allein es wird ihr an freier Entwickelung fehlen, sie wird in ihrem Wuchs frühzeitig kümmern und stocken, und sie wird nie erreichen, daß man von ihr sage: es walte in ihr etwas, das fähig sei, uns in Erstaunen zu setzen.] (FA, 2, 12: 599)

The oak tree will flourish only if its conditions are mixed. It must have enough nourishment to grow, and it must also engage in some trials: "a century's struggle with the elements makes it strong and powerful, so that, at its full growth, its presence inspires us with astonishment and admiration" (18 April 1827) [ein hundertjähriger Kampf mit den Elementen macht sie stark und mächtig, so daß, nach vollendentem Wuchs ihre Gegenwart uns Erstaunen und Bewunderung einflößt] (FA, 2, 12: 599). Beauty is therefore related to a balance or reconciliation of opposing elements. And while many of the factors that lead to the growth of a beautiful oak tree are outside of its control (the conditions of the soil, the proximity of its other neighbors, the climate, etc.), the tree also actively participates in the entire process. It, like many of the other natural entities in Goethe's scientific works, has a will that expresses a tendency (Tendenz), drive (Trieb), and striving (Bestreben). It actively attempts to flourish within the given scope of its "budget."

Goethe addresses similar issues of balance in his essay "The Extent to which the Idea 'Beauty Is Perfection in Combination with Freedom' May Be Applied to Living Organisms" ("In wiefern die Idee: Schönheit sei Vollkommenheit mit Freiheit, auf organische Naturen angewendet werden könne," 1794). This essay, written shortly before his essays on comparative anatomy (although not published in his lifetime), directly addresses the relationship between the principle of compensation and beauty. An animal is beautiful only if it has properly balanced its anatomical budget. He places such high value upon beauty not simply because he values outward appearances, but because he closely associates beauty with freedom. Only a well-proportioned and balanced individual can find freedom for its life because only such an individual will have the power most fully to execute its own will.

Goethe calls an animal perfect when it is so formed "that it may enjoy its existence, and maintain and propagate itself" (12: 22) [daß sie ihres Daseins genießen, dasselbe erhalten und fortpflanzen können, und in diesem Sinn ist alles Lebendige vollkommen zu nennen] (FA, 1, 24: 219). However, some animals, according to Goethe, are "more perfect" than others. He considers an animal less than perfect if it is unbalanced in some way: "limitation of organic nature to a single purpose will produce a preponderance of one or another of its members, rendering the free use of one or another of its members difficult" (12: 22) [denn durch die Beschränktheit der organischen Natur auf Einen Zweck wird das Übergewicht eines und des andern Glieds bewirkt, so daß dadurch der willkürliche Gebrauch der übrigen Glieder gehindert werden muß] (FA, 24: 219). The budget principle takes its toll on some animals, like the mole, which is perfect but ugly because "its form permits only a few, limited actions" (12: 21) [weil seine Gestalt ihm nur wenige und beschränkte Handlungen erlaubt und das Übergewicht gewisser Teile ihn ganz unförmlich macht] (FA, 1, 24: 219). The mole's very narrow form acts as a severe limitation to its freedom of movement and leaves one with an impression of disproportion.

If an animal is to be beautiful, it must be so proportioned that it is able to have enough strength to undertake "capricious actions which are somewhat without purpose" (*12: 22) [willkürliche gewissermaßen zwecklose Handlungen] (FA, 1, 24: 220). Goethe separates the idea of beauty from issues of survival and explains that the animal must have a kind of budget that is not too constrained for it to be beautiful. He uses the term *proportion* but states that this notion is outside of mathematical relationships (FA, 1, 24: 220). Proportion consists of the relation of parts to function as opposed to parts to other parts. If an animal is to be beautiful, its limbs need to be proportioned so

that none hinders the action of another; compulsion and need are entirely hidden from my sight by a perfect balance so that the animal seems free to act and work just as it chooses. We may recall the sight of a horse using its limbs in freedom. (12: 22)

[daß keins das andere an seiner Wirkung hindert, ja daß vielmehr durch ein vollkommenes Gleichgewicht derselbigen Notwendigkeit und Bedürfnis versteckt, vor meinen Augen gänzlich verborgen worden, so daß das Tier nur nach freier

Willkür zu handeln und zu wirken scheint. Man erinnere sich eines Pferdes das man in Freiheit seiner Glieder gebrauchen sehen.] (FA, I, 24: 220)

A beautiful animal has a body proportioned so that the organism can do a variety of activities. Only an animal with a properly balanced budget (i.e., only if it does not so favor one feature as to inhibit its ability to move freely) could be beautiful.

Goethe then turns to the theme of motion and rest. Beauty is related to the scientific enterprise of capturing movement at rest. Beauty is most evident when an animal is at rest but demonstrates an ability or the potentiality at any moment to move.[43] The animal must appear to have free use of its limbs according to its wishes, but it must be at rest:

Thus we see that beauty actually calls for *repose* together with *strength, inaction* together with *power.*

If the notion of asserting the power of a body or some limb is too closely associated with the being's physical existence, the spirit of the beautiful seems to take flight immediately: the ancients depicted even their lions in the greatest degree of repose and neutrality in order to draw forth the feeling which we grasp as beauty. (12: 23, original emphasis)

[Man sieht also daß bei der Schönheit *Ruhe* mit *Kraft, Untätigkeit* mit *Vermögen* eigentlich in Anschlag komme.]

[Ist bei einem Körper oder bei einem Gliede desselben der Gedanke von Kraftäußerung zu nahe mit dem Dasein verknüpft; so scheint der Genius des Schönen uns sogleich zu entfliehen, daher bildeten die Alten selbst ihre Löwen in dem höchsten Grade von Ruhe und Gleichgültigkeit, um unser Gefühl, mit dem wir Schönheit umfassen, auch hier anzulocken.] (FA, I, 24: 221, original emphasis)

His characterization of beauty is similar to his scientific method: the type or cloud nomenclature fixes nature for a moment, but that moment always is precariously close to changing as nature itself changes. Art, as the scientific method, fixes a form, but then requires one to recall the mobility of form as well.

When Goethe discusses human aspects of beauty, the criteria echo those of natural beauty. As previously seen, however, while human beings fit into a similar rubric as nature at large, they do so in an intensified state. Human beings are "almost free of the fetters of animality" [von den Fesseln der Tierheit beinahe entbunden], as their limbs are "in a delicate state of subordination and coordination, governed by will more than those of any other animal" (*12: 22–23) [mehr als die Glieder ir-

gend eines andern Tieres dem Wollen unterworfen] (FA, 1, 24: 220).
Goethe suggests that this bodily control, as much as language (logos) itself, raises human beings above other animals. He rates the human ability to speak through our limbs (gestures) as being equal to our ability to
use language (FA, 1, 24: 221). It is important to note that, while Goethe
ascribes a greater amount of will to human beings, the very statement
that he makes implies that animals too have a certain amount of control
as well: it is a matter of degree.

Beauty arises in any individual, whether human or nonhuman, as part
of a long process of compensation. The individual must first have developed a physical body that enables it to strive toward the free actions that
Goethe's definition of beauty requires. This sense of balance is not simply a formula for judging the physical organization and the exterior form
of animals. He states that developing the concept of a beautiful human
being requires a great deal of study. He links this project to an understanding of human freedom. Human balance goes beyond the proportion of the limbs or the free use of them. Human beings must balance
their spiritual characteristics in such a way as to enable freedom of both
thought and movement.

Goethe addresses the issue of achieving personal balance in several of
his essays. In "History of the Manuscript" ("Schicksal der Handschrift,"
1817), he further discusses human character using the terms of balance.
According to Goethe, all energetic, talented individuals are able to interest themselves in opposing tasks. After giving several specific examples
of individuals who are multifaceted, he concludes with an observation
that he believes is so true that it has become commonplace:

For the past several years we have heard repeated to the point of irritation the
eternal truth that a man's life is composed of both pleasure and work, and that
only the man who can maintain a balance between the two deserves to be called
wise and happy: for instinctively every person aspires to the exact opposite of his
nature, in order that he may achieve a perfect whole. (Mueller, 170)

[Seit mehreren Jahren wird uns zum Überdruß die ewige Wahrheit wiederholt,
daß das Menschenleben aus Ernst und Spiel zusammen gesetzt sei, und daß der
Weiseste und Glücklichste nur derjenige genannt zu werden verdiene, der sich
zwischen beiden im Gleichgewicht zu bewegen verstehe, denn auch ungeregelt
wünscht ein jeder das Entgegengesetzte von sich selbst, um das Ganze zu haben.]
(FA, 1, 24: 417–18)

Like the eye, human beings, when presented with one aspect, crave its opposite in order to have a whole. Goethe, in essence, takes quite seriously the notion of a well-rounded individual. He makes a very similar point in his 1824 essay "Ernst Stiedenroth," except here the focus is on the negative consequences of not achieving balance—of favoring one personality trait at the expense of another:

We are well enough aware that some skill, some ability, usually predominates in the character of each human being. This leads necessarily to one-sided thinking since man knows the world only through himself, and thus has the naive arrogance to believe that the world is constructed by him and for his sake. It follows that he puts his special skills in the foreground, while seeking to reject those he lacks, to banish them from his own totality. As a correction, he needs to develop all the manifestations of human character—sensuality and reason, imagination and common sense—into a coherent whole, no matter which quality predominates in him. If he fails to do so, he will labor on under his painful limitations without ever understanding why he has so many stubborn enemies, why he sometimes meets even himself as an enemy. (12: 45–46)

[Recht gut wissen wir, daß in einzelnen menschlichen Naturen gewöhnlich ein Übergewicht irgend eines Vermögens, einer Fähigkeit sich hervortut und daß daraus Einseitigkeiten der Vorstellungsart notwendig entspringen, indem der Mensch die Welt nur durch sich kennt und also, naiv anmaßlich, die Welt durch ihn und um seinetwillen aufgebaut glaubt. Daher kommt denn daß er seine Haupt-Fähigkeiten an die Spitze des Ganzen setzt und was an ihm das Mindere sich findet, ganz und gar ableugnen und aus seiner eignen Totalität hinausstoßen möchte. Wer nicht überzeugt ist, daß er alle Manifestationen des menschlichen Wesens, *Sinnlichkeit* und *Vernunft, Einbildungskraft* und *Verstand,* zu einer entschiedenen Einheit ausbilden müsse, welche von diesen Eigenschaften auch bei ihm die vorwaltende sei, der wird sich in einer unerfreulichen Beschränkung immerfort abquälen und niemals begreifen, warum er so viele hartnäckige Gegner hat, und warum er sich selbst sogar manchmal als augenblicklicher Gegner aufstößt.] (FA, I, 24: 615, original emphasis)

While such imbalances may lead to all kinds of problems, it is important to note that human beings have the power to correct any imbalances. We are free, according to Goethe, to transform our own characters to achieve balance. Where individual species of animals may take generations to correct the predominance of one feature, human beings may, as long as they have self-awareness, correct their own imbalances. It is also interesting to note that in addressing some of our possible imbalances, a type of teleology tops the list. People must avoid seeing nature as creating things for their own, "higher" ends. A balanced and well-rounded human being, therefore, sees himself or herself as a part of nature and its processes and not above it. Human imbalances, moreover, are equally likely to tend toward the rational as the irrational. Goethe's vision of a free human being is not a highly rational one, but one who is able to find an equilibrium between the rational and passionate. If human beings do not strive to balance their various features against one another, then they will face greater limitations and lose their ability to act freely. Human beings are most free when they have balanced their own characteristics so that one aspect of their personalities does not predominate and inhibit freedom of thought, feeling, or action.

Similarly, in his essay on beauty and freedom, Goethe implies that human freedom, or the "crown of human perfection," arises if one balances his or her own spiritual characteristics. He also hopes that his concept of beauty will eventually be linked with more spiritual principles. "[E]specially interesting" (höchst interessant) would be a discussion on "how limitation and specialization could appear without impairing freedom" (12: 23) [wie man Beschränkung und Determination aufs besondere, ohne der Freiheit zu schaden könne erscheinen lassen] (FA, 1, 24: 221). Such a discussion, he notes, if it is to be helpful to both friends of nature and art, would have to be based upon anatomy and physiology (FA, 1, 24: 221–22). While "it is not easy to imagine a form of discourse suitable for the presentation of such a varied and wondrous whole" (12: 23) [allein zur Darstellung eines so mannigfaltigen und so wunderbaren Ganzen hält es sehr schwer sich die Möglichkeit der Form eines angemessenen Vortrags zu denken] (FA, 1, 24: 222), Goethe appears to have discovered a suitable form of discourse with his own type and the principle of compensation. He began his work on his comparative anatomy essays— where he outlines his type and the principle of compensation—within a year after writing the essay on beauty.[44]

Finally, if we briefly turn to one of Goethe's most famous accounts of his own artistic process, the sonnet "Nature and Art" ("Natur und Kunst," 1807), we can further witness the role that he ascribed to compensation within his own artistry. The poem "The Metamorphosis of Animals," generally states that the principle of compensation is present in works of art. "Nature and Art" provides a more detailed account of

how this principle operates. The first part of the poem addresses the role
of seemingly opposite forces:

> Nature, it seems, must always clash with Art,
> And yet, before we know it, both are one;
> I too have learned: Their enmity is none,
> Since each compels me, and in equal part.
> (1: 165)

> [Natur und Kunst sie scheinen sich zu fliehen,
> Und haben sich, eh' man es denkt, gefunden;
> Der Widerwille ist auch mir verschwunden,
> Und beide scheinen gleich mich anzuziehen.]
> (FA, I, 2: 838)

The poet initially views art and nature as opposing entities that cancel
each other out. In the context of the poem, art is seen as a constricting
element. It requires hours of labor by the artist to bring forth his or her
product. Nature, conversely, sets the heart free from the restrictions of
work:

> Hard, honest work counts most! And once we start
> To measure out the hours and never shun
> Art's daily labor till our task is done,
> Freely again may Nature move the heart.
> (1: 165)

> [Es gilt wohl nur ein redliches Bemühen!
> Und wenn wir erst in abgemess'nen Stunden
> Mit Geist und Fleiß uns an die Kunst gebunden,
> Mag frei Natur im Herzen wieder glühen.]
> (FA, I, 2: 839)

The poet soon discovers that these two opposing forces do not in fact
act in a negating manner, but in a complementary one. It is not enough
for an artist's heart freely to be moved and inspired by nature, although
this inspiration is a necessary component to his art:

> So too all growth and ripening of the mind:
> To the pure heights of ultimate consummation
> In vain the unbound spirit seeks to flee.

Who seeks great gain leaves easy gain behind.
None proves a master but by limitation
And only law can give us liberty.
(1: 165)

[So ist's mit aller Bildung auch beschaffen:
Vergebens werden ungebundne Geister
Nach der Vollendung reiner Höhe streben.]

[Wer Großes will muß sich zusammenraffen;
In der Beschränkung zeigt sich erst der Meister,
Und das Gesetz nur kann uns Freiheit geben.]
(FA, I, 2: 839)

In these lines, Goethe draws upon a theme that he repeatedly uses in his scientific works. Creativity, to reach its highest state, requires limitation. Conversely, too many limits (as represented in the poem by the labors of art without the freedom that nature provides) would cripple the artist's ability to create. The artist needs to have freedom and restrictions to create the highest products with the greatest ease of expression. Goethe's natural principle of compensation seems very much to be at play within the poem. The artist, like the evolving whale, must learn how to balance freedom against limits. He or she must learn to work within certain restrictions and limitations in order to create the most perfect forms.[45]

❧

This chapter has examined the configurations of organic and inorganic forms according to Goethe's principle of compensation. Animals, human beings, plants, and even clouds reach their highest aesthetic and functional forms when they manage to balance their will against physical obstacles. The next chapter, which discusses Goethe's conception of reproduction and gender, demonstrates just how difficult achieving this balance may be. The process of reproduction illustrates that often an organism does not have one will, but several competing ones that are vying for control.

4 COMPETITION, REPRODUCTION, AND GENDER

 GOETHE'S POEM "Amyntas" portrays the relationship between the sexes as highly sadomasochistic. Amyntas, the speaker of the poem, tells a story about a tree that serves as metaphor for his own destructive relationship with his beloved.[1] An apple tree no longer bears fruit because an ivy is slowly choking it to death: the ivy, the traitor (Verräterin), enjoys (genießt) what the apple tree needs [Nahrung nimmt sie von mir; was ich bedürfte, genießt sie] (FA, 1, 2:195). Where the image of the vine and the tree traditionally represents fertility (brought about by the structural dependence of the "feminine" vine upon the "masculine" tree),[2] the poem changes the metaphor to represent a destructive relationship. The ivy has surrounded the tree and sucks the lifeblood and very soul from it (saugt sie das Mark, sauget die Seele mir aus).[3] The tree, while openly recognizing the ivy as a dangerous guest (der gefährliche Gast), rejoices in the bondage that will lead to its death (freue der Fesseln, / Freue des tötenden Schmucks, fremder Umlaubung mich nur).

Goethe's revision of a traditional image shows how easy it may be to move from a balanced and mutually beneficial relationship to a destructive one. His scientific works, especially his botanical treatises, deal extensively with the struggle for power between the masculine and the feminine. His discussion of the masculine and feminine is unusual in that it does not focus on sexual organs, but rather upon certain characteristics that may be totally unrelated to particular body parts. A plant may, for example, contain both sexual organs but display predominantly "feminine" characteristics, or it may display "masculine" characteristics and yet have no sexual organs at all, etc. At the focal point of Goethe's study of these forces is their competitive drives. These two forces are competitive due to their similarities: both wish to reproduce in their own way, both strive to exert their will, both are equally necessary to an individual organism, and both represent heightened/intensified development. And, as sexual organs are not at the forefront of his discussion, neither is sexual reproduction. Although the masculine and the feminine play very large roles in Goethe's principle of reproduction, sexual reproduction is by no means the only method of reproduction of which nature is capable. Chapter 2 already discussed how the moment of sexual union in a plant represents a kind of created thing apart from the seed. In other scientific essays, Goethe discusses other possibilities of reproduction and broadly extends the notion of reproduction to include any kind of growth. He is able to do so because he does not view an organism as a solitary individual, but as a community comprised of many parts. The plant therefore, is a community of leaves, eyes, buds, etc., whereas an animal's organs or discrete parts (such as the numerous shells of a barnacle or the organ of the eye) comprise its community. In his view, not only do whole organisms strive to reproduce their kind, but individual parts within a single organism constantly strive to reproduce as well. Because resources are often scarce, this struggle may be fiercely competitive. Moreover, an overreproduction of one part—such as when a plant produces too many eyes—may cause the collapse or death of the organism.

Descartes, in his writings on reproduction, had tried to account even for propagation and organic development through mechanistic principles: "It is no less natural for a clock, made of a certain number of wheels, to indicate the hours, than for a tree born from a certain seed, to produce a particular fruit" (*De la formation de l'animal*, quoted in Roe, 4). Although Goethe was not alone in questioning the viability of a wholly mechanistic explanation for reproduction or life, the mechanistic argument to explain life was at the center of the argument between the preformationists and epigenesists—an argument that continued throughout Goethe's life (Farley, 30–31) and occupied his thoughts. Buffon had tried to explain reproduction through forces analogous to gravity and had considered semen to be a type of machine (Roe, 16–17, 18).[4] Haller, the most famous proponent of preformation, was a committed Newtonian mechanist (Lenoir, "Eternal," 18; Roe, 2).[5] And while many German biologists in Goethe's time argued for epigenesis, they often did so from a perspective that focused upon nature's reactive qualities. As Farley reports, epigenesists of the time "tended to view fecundation in essentially chemical terms: two fluids come together, react, and produce a new being" (30).[6] Goethe approached his reproductive studies from a different, and much more vitalistic, perspective. For Goethe, each part of an organism played an active role in the growth and development of the whole. At any time, a single part of an organism may so exert its will as to reproduce itself and therefore change the course of development of the whole organism. His discussions of reproduction therefore were not limited to nature's more regular manifestations, but addressed the irregular as well.

The Coming-to-be of Parts and Wholes

Goethe's conception of the organism emphasizes the multiplicity and the interrelationship of its parts. Nature is so complicated that we should never assume that we are studying a homogeneous whole; we are always looking at a complex entity comprised of many individual parts. Goethe's philosophy of reproduction focuses upon nature's creative acts, including the replication of the smallest part, the formation of a flower, or the birth of an entire animal.

In the poem "Epirrhema" (1819), Goethe describes how a single organism is also a community of individuals. The poem begins by stressing the unity of nature. Any entity within nature is related in an analogous way to all other entities. In studying natural entities, however, we are to look beyond single organisms or things in order to discover nature's multiplicity of forms:

> You must, when contemplating nature,
> Attend to this, in each and every feature:
> There's nought outside and nought within,
> For she is inside out and outside in.
> Thus will you grasp, with no delay,
> The holy secret, clear as day.
>
> ❧
>
> Joy in true semblance take, in any
> Earnest play:
> No living thing is One, I say,
> But always Many.
> (1: 159)

> [Müsset im Naturbetrachten
> Immer eins wie alles achten;
> Nichts ist drinnen, nichts ist draußen:
> Denn was innen das ist außen.

So ergreifet ohne Säumnis
Heilig öffentlich Geheimnis.

❧

Freuet euch des wahren Scheins,
Euch des ernsten Spieles:
Kein Lebendiges ist ein Eins,
Immer ist's ein Vieles.]
(FA, I, 2: 498)

He talks of the "earnest play" of nature because, on the one hand, the dynamic interaction of nature's parts may appear to be a game. Their movements may appear capricious and spontaneous. Within *Elective Affinities,* the characters also see the interaction of chemical elements as a game, in which chemicals play out a variety of social structures, from friendships to class distinctions. On the other hand, the game is quite earnest because understanding the dynamic structures of an entity enables us to comprehend its activities. Had the character Charlotte remembered that some chemicals find each other so attractive that they are impossible to separate once they have joined, she might not have attempted to keep Ottilie and Eduard apart. Her attempts, in the end, contribute to three deaths in the novel. Similarly, in the scientific sphere, understanding why one part of an organism dominates over another helps one to understand the vital energies behind its formation. It also illustrates how precarious any natural order may be. Due to the multiplicity of wills of the various members of the community, hierarchical order may break down at any time.

The theme of the many within the one is prevalent throughout Goethe's scientific texts. Within these works, he presents a somewhat unusual conception of an organism. He argues that any individual organism is not really a unified whole but a complex community of parts. Not only do single organisms reproduce, but each organism itself contains many parts that engage in their own reproductive processes:

No living thing is unitary in nature; every such thing is a plurality. Even the organism which appears to us as individual exists as a collection of independent living entities. Although alike in idea and predisposition, these entities, as they materialize, grow to become alike or similar, unalike or dissimilar. In part these entities are joined from the outset, in part they find their way together to form a union. They diverge and then seek each other again; everywhere and in every way they thus work to produce a chain of creation without end. ("The Purpose Set Forth," 12: 64)

[Jedes Lebendige ist kein Einzelnes, sondern eine Mehrheit; selbst insofern es uns als Individuum erscheint, bleibt es doch eine Versammlung von lebendigen selbständigen Wesen, die der Idee, der Anlage nach, gleich sind, in der Erscheinung aber gleich oder ähnlich, ungleich oder unähnlich werden können. Diese Wesen sind teils ursprünglich schon verbunden, teils finden und vereinigen sie sich. Sie entzweien sich und suchen sich wieder und bewirken so eine unendliche Produktion auf alle Weise und nach allen Seiten.] ("Die Absicht eingeleitet," 1817, FA, I, 24: 392)

Each part is viewed as a single entity or individual that is capable of exerting its own influence upon the community.[7] The last chapter already addressed some of the consequences of emphasizing one body part over another. Goethe's discussion of an organic community, however, even further emphasizes his creative notion of nature, because it is not simply the whole organism that expresses a will, but many of its parts as well.

Although Goethe focuses upon the various parts of a single entity, his approach is antithetical to one that dissects an organism to discuss its parts.[8] Rather than seeking to establish mechanical explanations for the relationship of the parts to the whole, he stresses the dynamic interactions of the parts. The language in the text is reminiscent of Goethe's creative notion of polarity and intensification, where new entities are formed through separations and reunions. The plant, as already discussed in chapter 2, participates in this kind of process when it develops two different sexual organs that then reunite and form the seed. Sexual union of plant organs, however, is not the only example of the interaction of parts or polarities within one community. Within the community of parts, numerous different individual parts may reproduce themselves:

Although a plant or tree seems to be an individual organism, it undeniably consists only of separate parts which are alike and similar to one another and to the whole. How many plants are propagated by runners! In the least variety of fruit tree the eye puts forth a twig which in turn produces many identical eyes; propagation through seeds is carried out in the same fashion. This propagation occurs through the development of innumerable identical individuals out of the womb of the mother plant. ("The Purpose Set Forth," 12: 64)

[Daß eine Pflanze, ja ein Baum, die uns doch als Individuum erscheinen, aus lauter Einzelheiten bestehn, die sich untereinander und dem Ganzen gleich und

ähnlich sind, daran ist wohl kein Zweifel. Wie viele Pflanzen werden durch Ab-
senker fortgepflanzt. Das Auge der letzten Varietät eines Obstbaumes treibt
einen Zweig, der wieder eine Anzahl gleicher Augen hervorbringt; und auf eben
diesem Wege geht die Fortpflanzung durch Samen vor sich. Sie ist die Entwick-
lung einer unzähligen Menge gleicher Individuen aus dem Schoße der
Mutterpflanze.] ("Die Absicht eingeleitet," FA, I, 24: 393)

This community of parts maintains its individuality and propagates its
kind through a variety of methods, from asexual (runners) to heterosexual reproduction (seed). This community is a complex one, where even
seeming unities or wholes, such as the seed, may be further broken down
into even more complex parts. Goethe's description of the seed illustrates that the plant, even in its most simple and early stage, is a community of parts The seed represents, *in potens,* the extraordinary desire of
nature's individual parts to exert their own influence. Nature in the
small scale is identical to nature in the large scale. Its parts struggle to
exist and thrive.

In his description of the seed, Goethe, but for the repeated emphasis
upon the changing and malleable "life of nature" (FA, I, 24: 394), seems
to endorse a preformationist view, where the embryo appears to contain
a miniature version of the adult (FA, I, 24: 393–94). However, the differences between Goethe's conception of reproduction and that of preformation are profound. He does not envision that the shape of the adult is
predetermined at this stage. The "fetus" of the plant, the seed, already
contains a community of different parts. At this incipient stage, it is impossible to tell which parts of the community will dominate, i.e.,
whether the plant will progress regularly and continue onto sexual union
or whether another part will take control and force the plant to progress
along a different path.[9] Similarly, an eye of a plant, which also reproduces
the plant if planted or grafted, is a community of parts ("Metamorphosis," no. 90). The "secret" of propagation has to do with the desire of individual parts to exert their power and influence on the development of
the plant and, while the parts may be similar, once one has gained control over the others, differences are more likely to emerge.

Goethe expands the definition of procreation to include any kind of
growth so that a plant engages in reproductive activity any time it grows
or expands.[10] Growth becomes a form of reproduction because the individual part (e.g., the leaf) is reproduced as the plant grows. If a plant is

overwatered, the sexual parts of the plant will not gain hierarchical control of the community, but the nonintensified organ of the leaf will instead continue to replicate its parts and reproduce vegetatively in leafy growth. The sexual organs will never even form. Such a plant exhibits the more "democratic" pattern of growth, i.e., a continual reproduction of equal parts rather than the more hierarchical reproduction enabled through the intensified parts of the masculine and feminine organs. In other words, he envisions reproduction on an individual and democratic scale (e.g., the leaf) as well as a communal and more hierarchical scale (the entire community of the plant reproducing itself in the seed). Accordingly, he sets forth two categories that describe a plant's ability to reproduce: vegetatively (which he links in his "Spiral Tendency" essay to masculine characteristics) and reproductively (which he links in the "Spiral Tendency" to feminine ones).

These two modes of reproduction are significant because they demonstrate nature's creative power. Nature strives to create at every opportunity, and we can observe this desire not only in heterosexual reproduction, but throughout all organic growth. Goethe calls the vegetative reproduction successive and identifies it as sequential ("Metamorphosis," no. 113). It is the more simple, slow, and primary method of reproduction (no. 86). We witness this type of reproduction when a plant grows from node to node and expands its growth through the production of leaves. The reproductive or "feminine" mode, in contrast, is simultaneous and represents the height of the greatest contraction and specialization of the plant as it reproduces through sexual union. It is further interesting because it, like the masculine, may replicate on its own. For example, the eyes of a plant are considered to be a part of the feminine mode (FA, 1, 24: 787), and eyes are just as capable of reproducing a plant as is a seed ("Metamorphosis," no. 87). Like the organs of our own eyes, the eyes of a plant represent a potential and a desire for wholeness. Once separated from the mother plant, they have the impetus to create a new whole for themselves.

The relationship between the growth and reproductive methods of propagation is complex. Some plants, such as the perfoliate or double rose, participate in "masculine" reproduction by producing flower after flower instead of allowing complete sexual organs to form ("Metamorphosis," nos. 103–4). Conversely, "feminine" reproduction may be fostered at the expense of "masculine." If a plant is underwatered, it will so quickly strive toward sexual or simultaneous reproduction that masculine, leafy growth will be sacrificed. Goethe's literary heroine, Ottilie, represents this type of growth. Throughout the novel, she is compared to a progressive plant. In the end, by starving herself to death, she cuts off the possibility of either a procreative life with her lover, Eduard, or a productive one as teacher.[11]

Masculine and feminine characteristics or parts of plants, then, like numerous other parts of an organism, exert an urge or will to exist and to promote the flourishing of their own parts. That they have this same desire whether acting alone or in concert illustrates how basic the creative drive is to nature's various parts. Moreover, because the masculine and the feminine can reproduce on their own, Goethe's conception of individual roles of the genders differs significantly from that of his contemporaries. He emphasizes the creative capabilities of both. The feminine is not a passive participant in generation, nor is it even an issue for Goethe, as it was for many of his contemporaries (Farley, 16–30; Pinto-Correia, 16–135; Roe, 1), whether the masculine or the feminine contributes the matter or the form to the offspring.[12]

Goethe's focus on the striving of whole organisms and their parts demonstrates the centrality of nature's creative desire. The creation myth of "Reunion" described the yearning of polar halves for each other. In Goethe's botanical works, in addition to the desire of nature's parts to reunite (as in the examples of sexual reproduction), its parts may also have creative desires apart from sexual ones. These desires are at times disruptive, destructive, and competitive. Through these very interruptions of nature's more regular laws, however, nature may reproduce itself and create new forms. The study of its competitive battles and abnormalities, therefore, highlights its creative energies.

Natural Insurrections

Goethe's natural studies take on a decidedly political tone as he describes the relationships of nature's parts to the whole.[13] Sometimes a hierarchy reigns within a natural organism, while at other times chaos results when some parts seek to overthrow the order of the whole. The principle of compensation demonstrates that the parts of any organism are so closely related to one another that a change in one will influence the others. By more closely examining the relationship of these parts, one becomes aware that their relationship is not simply characterized by the give and take of the principle of compensation. Rather, at any time, a revolution among the parts may drastically change the structure of the whole. The political structure of these communities is variously a strict hierarchy, a democracy, or anarchy, depending upon which part gains ascendancy.

In Goethe's account, the process of growth and reproduction is not always a smooth one. Sometimes one part attempts to exert its creative will at the expense of other members of the community:

Every leaf, every eye, has in itself the right to become a tree. It is only the ruling health of the stem that prevents the leaves and buds from attaining that state. We cannot repeat often enough that every organization unites various active parts. In the present instance, we observe that the stem is usually round or at least may be regarded as round within. And it is precisely the roundness, in its singleness, that holds asunder the individual parts such as leaves and buds, and allows them to ascend in ordered sequence to regular development until they blossom and bear fruit. But if such a plant entelechy is checked, if not indeed done away with completely, the middle loses its governing power, the periphery contracts, and each individual, striving part now exercises its own special right. ("Later Studies and Collections," *Mueller translation, 100)

[Jedes Blatt, jedes Auge an sich hat das Recht ein Baum zu sein; daß sie dazu nicht gelangen bändigt sie die herrschende Gesundheit des Stengels, des Stammes. Man wiederholt nicht oft genug, daß jede Organisation mancherlei Lebendiges vereinige. Schauen wir im gegenwärtigen Falle den Stengel an, dieser ist gewöhnlich rund oder von innen aus für rund zu achten. Eben diese Ründe nun hält als Einheit die Einzelnheiten der Blätter, der Augen aus einander und läßt sie, in geordneter Nachfolge, aufsteigen zu regelmäßiger Entwickelung bis zur Blüte und Frucht. Wird nun eine solche Pflanzen-Entelechie gelähmt, wo nicht aufgehoben, so verliert die Mitte ihre gesetzgebende Gewalt, die Peripherie drängt sich zusammen und jedes Einzelnstrebende übt nun sein besonderes Recht aus.] ("Nacharbeiten und Sammlungen," 1820, FA, I, 24: 464–65)

Goethe's emphasis on the activity of the separate parts is reminiscent of Lucretian atomism. Every part may act independently and influence the ultimate outcome of the whole or form an entirely new whole. The independent activity of each part is generally held in check by the hierarchical control of the plant's regular pattern of growth. If that structure is weakened, each individual part may exercise its own "rights." The tree, for example, may no longer grow in a straight, upward pattern, but begin to grow in curves and crooks as the eyes and nodes multiply more rapidly than its other parts. And although Goethe might seem to endorse a more Aristotelian position by referring to the regular process as *entelechy*, it is clear from the preceding passages that he did not see this type of growth as exclusive. Instead, he warns against regarding the abnormal as diseased or pathological and against using such terms as misdevelopment, malformation, crippling, and stunting (Mißentwickelung, Mißbildung, Verkrüppelung, Verkümmerung) to describe it.

Goethe instructs us to avoid such terms for two reasons. First, these "abnormal" growths are just as much a part of nature as the more normal ones. Second, nature, in these cases, exerts the greatest freedom (mit höchster Freiheit wirkend) and may discover new types of perfections (FA, 1, 24: 462–63). He turns to the example of the perfoliate or double rose as an organism that disrupts entelechy. It is "monstrous" because the step-by-step pattern of the plant's growth has been disrupted (stems grow in the flower, preventing mature sexual organs to form, but thereby enable additional flowers to form). Because of this disruption, however, it is also more beautiful and more fragrant than a flower created through the regular process (FA, 1, 24: 462). Goethean nature, unlike Aristotelian or Cartesian nature, may create new patterns and goals: "Nature oversteps the boundary she has set, but attains thereby a different kind of perfection; thus we would do well to defer the use of negative terminology as long as possible" (Mueller translation, 99) [Die Natur überschreitet die Grenze die sie sich selbst gesetzt hat, aber sie erreicht dadurch eine andere Vollkommenheit, deswegen wir wohltun uns hier so spät als möglich negativer Ausdrücke zu bedienen] (FA, 1, 24: 462). By not expecting nature to be regular in all of its activities, scientists are much more likely to become aware of the positive aspects of its irregularities and to see the dynamic possibilities of natural change at any stage.

Goethe argues that the "abnormal" and the "irregular" are just as much a part of nature's "order" as its more "regular" processes. The more we study irregular forms, the more we become aware that nature's laws are themselves at times irregular and dynamic. Goethe's characterization of the political relationship of plant parts is closely connected to his natural philosophy as a whole. While a hierarchy of parts may generally enforce a regular pattern of development, one may never know at what point nature may depart from its regular course or when the regular and the irregular may become intertwined. One therefore cannot either assume regular, formulaic patterns or assign negative values to irregular phenomena:

Nature fashions normally when she subjects innumerable particulars to a rule, defines and delimits them. Conversely, the phenomena are abnormal when the particulars carry the day, emerging in an arbitrary, indeed apparently accidental way. However, since both are closely related, and since the same spirit animates the regular and the irregular as well, an oscillation occurs, formation and transformation forever alternating, so that the abnormal seems to become normal and the normal, abnormal. ("Later Studies and Collections," Mueller translation, 98)

[Die Natur bildet normal, wenn sie unzähligen Einzelheiten die Regel gibt, sie bestimmt und bedingt; abnorm aber sind die Erscheinungen, wenn die Einzelnheiten obsiegen und auf eine willkürliche, ja zufällig scheinende Weise sich hervortun. Weil aber beides nah zusammen verwandt und, sowohl das Geregelte als Regellose, von Einem Geiste belebt ist, so entsteht ein Schwanken zwischen Normalem und Abnormem, weil immer Bildung und Umbildung wechselt, so daß das Abnorme normal und das Normale abnorm zu werden scheint.] ("Nacharbeiten und Sammlungen," 1820, FA, I, 24: 462)

Goethe admonishes the natural observer to be careful with artificially constructed categories. He argues that it is too easy to dismiss nature's more dynamic creations as exceptions to the rule rather than part of a dynamic relationship. One must therefore study "abnormal" entities alongside more "normal" or regular natural productions because they are indeed related to one another: both are impelled by the same creative spirit.

Goethe suggests that neither the parts of an organism, nor the relationship of the parts to whole, may be explained simply by matter and motion. Where the proponents of Cartesian mechanism excluded vital characteristics and focused upon inert matter and motion (Farley, 16), Goethe focused upon vitality as the most telling aspect of organic life. Goethe's scientific works portray a nature that is constantly reproducing itself in numerous different parts. The political relationship of the parts to the whole largely determines which parts will reproduce and which will be suppressed. During his era's fascination with sexuality and reproduction,[14] his scientific works examine numerous methods of regeneration apart from heterosexual ones as well as acknowledge the fierce competition among the various elements that struggle to come into existence.

Turning now to an example in the animal world, it becomes even more evident how discordant the creative struggle may be. Nature is at its most ruthless when resources are scarce. This view of nature's competitive drives once again reflects Goethe's rejection of a mechanistic view of nature. Because competition arises among the parts of one organism as well as between different organisms, one must study the relationships of the different participants and their particular environmental circumstances to understand the outcomes of the competition. One of Goethe's most vivid and philosophically important discussions of the competition of parts striving to come into existence occurs in his essay "The Barnacles" ("Die Lepaden"). It is interesting to note that it was Charles Darwin's work with barnacles that enabled him to discover the importance of individual variability and "made him realize that the struggle for existence due to competition . . . was a phenomenon involving individuals and not species" (Mayr, 487). Barnacles proved to be such an interesting study for Darwin because their "individual variability" was so great, that he often doubted "whether two specimens were variants of a single species or two different species." This was especially true since their variability was not limited to their external appearance, but could be witnessed even in their internal organs (Mayr, 487). A generation earlier, Goethe's will-driven conception of nature made him look for competition in an even smaller sphere: he examined the competition among the parts within one individual.

The Barnacles and the State of Nature

"The Barnacles" (1824) is a very short essay that compares the growth patterns of two different kinds of barnacles: the *Lepas anatifera* (goose barnacle)[15] and the *Lepas polliceps*. Within the essay, Goethe focuses upon their differences in shell formation. The shells of *Lepas anatifera* grow regularly and symmetrically, whereas those of *Lepas polliceps* grow asymmetrically and irregularly. The two growth patterns become symbolic of nature's processes of growth and development. Like the plant, the barnacle is considered both a solitary being as well as a community comprised of numerous individual parts. The two different barnacles, however, represent different political structures, the goose barnacle an ordered and hierarchical one, whereas the *Lepas polliceps* a more chaotic and disordered one.

Goethe's discussion of the more regular barnacle concentrates on its growth patterns. The number of shells is always limited to five, their size is basically fixed, they appear in predetermined and regularly spaced distances from one another, and their spacing has direct correspondence to particular body parts (FA, 1, 24: 611). They illustrate nature's more ordered and rule-based aspects. Because the patterns of growth and the relationship of the parts are ruled by a hierarchy, one cannot see the individual desires of this barnacle's individual parts.

A study of these beings leads the natural observer into a dead end. No matter how long one observes them, they do not shed enlightenment about the inner workings of nature [Hierüber würde nun eine noch so lange Betrachtung der *Lepas anatifera* uns nicht weiter aufklären] (FA, 1, 24: 611). Their regularity hides from view the hidden forces of nature. Goethe advocates studying the more irregular barnacles to gain knowledge of nature and to awaken the deepest, most general convictions (die tiefsten allgemeinsten Überzeugungen). As the irregular plant unveils hidden aspects of the regular plant ("Metamorphosis," no. 7), the irregular barnacle illuminates principles that lie hidden in the more regular one (FA, 1, 24: 611). The individual shells of this barnacle much more clearly demonstrate nature's strong urge to exist and create.

A close examination of this barnacle demonstrates how each of its parts displays an intense, competitive desire to exist. Its outer skin is rough and covered with countless tiny round points that Goethe characterizes as sublime (erhaben). Each of these points is potentially a shell, and each awaits an opportunity to come into being. Goethe describes the growth of these shell points on the bottom and the top of the organism. Both growth patterns demonstrate central aspects of his natural philosophy. As soon as the bottom body or hose part of the barnacle expands, the little points are given an impetus (Antrieb) to become actual. Whereas the *Lepas anatifera* is limited to five shells, the number of shells in *Lepas polliceps* is practically without limit. The only limit to the number of shells is space. If the hose does not have room to expand, then no additional shells will form. As long as space exists for expansion, the number of shells will continue to grow. Nature will continue to create at every opportunity until it is forced to concede to some exterior limit.

Similarly, the growth of the shells on the top of these barnacles represents nature's heightened creativity within limits. Sometimes the main shells of the irregular barnacle do not grow large enough to cover or protect the body of the creature. Goethe here admires (bewundern) the business (Geschäftigkeit) of nature as it strives to solve this problem. It is able to compensate (ersetzen) for the deficiency of power (Mangel der Kraft) through the quantity of activity (Menge der Tätigkeit). Smaller shells will form, row upon row, to close the gap. The final row of shell points will appear to be a small string of pearl-like points around the border of the creature's opening. This final row of shell points will always be denied the passage from potentiality (Möglichkeit) to actuality (Wirklichkeit). Shells therefore will continue to form until the good of the creature requires them to stop.

One readily recognizes the principle of compensation in these passages. The barnacle creates its armor of shells within the limits of its environment on the bottom and its own well-being on the top. Goethe's description of this barnacle, however, probes more deeply into the competitive relationships of the parts within an individual than may be illustrated by his principle of compensation. In describing the relationship of each shell point to the others, he paints a picture of nature reminiscent of Hobbes. Each potential shell point struggles to become actual and achieves this goal at the expense of other shell points:

and here, by more exact observation, it seems as if every shell point rushes to consume the next one, to enlarge itself at the cost of the others, and does so in that moment before they are able to come into existence. An already formed, ever so small shell, cannot be eaten by its approaching neighbor—all things that have come into existence place themselves in an equilibrium against other existing things. And so one sees that the growth, which is regularly bound and lawlike in the goose barnacle [*Lepas anatifera*], is encouraged to freer advancement in the other barnacle [*Lepas polliceps*], where many an individual point usurps as much property and space as it can gain. (My translation)

[und hier, bei genauer Betrachtung, scheint es als wenn jeder Schalpunkt sich eile, die nächsten aufzuzehren, sich auf ihre Kosten zu vergrößen, und zwar in dem Augenblick ehe sie zum Werden gelangen. Eine schon gewordene noch so kleine Schale kann von einem herankommenden Nachbar nicht aufgespeist werden, alles Gewordene setzt sich mit einander ins Gleichgewicht. Und so sieht man das in der Entenmuschel [*Lepas anatifera*] regelmäßig gebundene, gesetzliche Wachstum, in der andern [*Lepas polliceps*] zum freiern Nachrücken aufgefordert, wo mancher einzelne Punkt so viel Besitz und Raum sich anmaßt als er nur gewinnen kann.] (FA, I, 24: 612)

Because the number of its shells is not limited, one is able to witness over and over again the fierce battle of the shells as they strive to come into being. Each individual tries its best to be born, and each individual will do whatever it can to assure its own position. Where the goose barnacle represents a nature that is hierarchically driven and well ordered, the Lepas polliceps symbolizes a fiercely competitive world. Nature's drive to create and procreate exists within the smallest parts of an organism. These small worlds or individual shell points follow one of the more general principles of nature—the drive toward existence. Once they exist, they no longer need "fear" the other shell points. They have won their right to exist and cannot be threatened by future shell points. Goethe similarly recalls nature's main drive in his "Problems" essay (1823) when he describes "the drive for specific character, the stubborn persistence of things which have finally attained reality" (12: 43) [den Spezifikationstrieb, das zähe Beharrlichkeitsvermögen dessen was einmal zur Wirklichkeit gekommen] (FA, I, 24: 582–83). One of nature's fiercest struggles involves the struggle of individual beings to come into existence. The birth of the individual shells is an example of how intensely each individual part of nature feels this drive. Each individual so wishes to be alive that it does whatever it can to exist.

Goethe argues, however, that these barnacles are not therefore governed by total chaos or confusion (Verwirrung). Rather, the more irregular barnacle will take on a more regularly patterned appearance if we consider each shell and each shell point as individuals—as small worlds (kleine Welten) that are competing against others for space. Viewed as a single entity, the barnacle appears to be in a state of confusion where shell crowds upon shell. If we consider each shell to be its own individual, each little world engages in its own individual struggle to exist, just as each whole organism does whatever it can to thrive.

Goethe writes that observing the coming-into-being of these points through a microscope would be one of most wonderful plays (eins der herrlichsten Schauspiele) that the friend of nature could ever wish to see. This particular play goes even further than those "other" productions involving color and clouds. Here one witnesses the coming-into-being of new worlds. These shell points participate in the same process of polar creation that Goethe describes within his autobiography. The pulse of creation and destruction symbolically attributed to Lucifer and the godhead comes alive. Even within the barnacle essay, he endows these competitive little beings with natural and religious symbolism:

Because according to my way of researching, knowing, and enjoying, I allow myself to hold only to symbols, so these creatures [*Lepas pollíceps*] belong to the sacred relics that in a fetishlike manner always stand before me. And they, through their odd structure, that striving toward the ruleless (irregular), always regulating themselves and therefore, in the small as well as large scale, make present to the senses godlike and humanlike nature. (My translation)

[Da ich nach meiner Art zu forschen, zu wissen, und zu genießen, mich nur an Symbole halten darf, so gehören diese Geschöpfe [*Lepas pollíceps*] zu den Heiligtümern welche fetischartig immer vor mir stehen und, durch ihr seltsames Gebilde, die nach dem Regellosen strebende, sich selbst immer regelnde und so im kleinsten wie im größten durchaus Gott- und menschenähnliche Natur sinnlich vergegenwärtigen.] (FA, I, 24: 612–13)

The irregular barnacles are more like human beings and God because they contain a determination to exert their will. These barnacles, like Faust, strive to reach the outermost limits of their existence. They are not governed by regular rules, but strive to break free as much as possible from patterned behavior. However, like Faust, they too in the end give a rule to themselves. Faust dies and ascends to heaven, not when he rests in a moment of satisfaction, but when he can imagine at what point such a moment might occur—at what point he might be able to give himself a limit to his own striving. This limit involves creating something that he believes will benefit human beings generally: a land reclamation project, i.e., a project that will expand the Lebensraum of human beings. Similarly, the irregular barnacles appear at first to exist ruthlessly as each shell strives to consume the others to gain the right to exist. The striving of the individual shell points is also subsumed by the good of the whole entity. The organism self-regulates its own growth on top so that it stops growing before it chokes itself by closing its own opening.

In each of the aforementioned cases, it is of central importance to treat organisms as community of parts. Sometimes that community has a hierarchical and predetermined structure as evidenced in the regular plant and barnacle. Other times, the structure is more fluid and represents a struggle for power. Although Goethe does not reject all natural laws, he warns that if one focuses exclusively upon the regular aspects of nature, one misses the more interesting and informative ones that demonstrate nature's creative impulses.

Goethe's accounts of plant and barnacle reproduction—where reproduction is broadly understood to represent any kind of growth or expansion—emphasize how parts of organisms strive to come into existence. His discussion of the masculine and feminine parts of organisms further reveals the inherent conflicts that exist within natural organisms as their individual parts strive to flourish. The battles between the masculine and the feminine tendencies are competitive because these tendencies are so similar.

The Natural Philosophy of Sex and Gender

In a fragment from his botanical notes, Goethe muses that one of the difficulties in gaining an understanding of sexual difference has been that we have considered the two main sexes separately and not as two parts of the same individual:

One cannot grasp the right concept of the two sexes if one does not imagine them in one individual. This sentence seems all too paradoxical because our concepts begin from human beings or from developed animals, and we therefore most easily distinguish the two sexes when we picture them on two individuals.
For this, the plants give us the best opportunity. (My translation)

[Man kann den rechten Begriff von den zwei Geschlechtern nicht fassen, wenn man sich solche nicht an einem Individuum vorstellt. Dieser Satz scheint allzu paradox zu sein, da unsere Begriffe sich vom Menschen oder von den ausgebildeten Tieren anfangen und wir eben dadurch am besten die beiden Geschlechter unterscheiden, daß wir sie an zwei Individuen wahrnehmen.
Hierzu geben uns die Pflanzen die beste Gelegenheit.] ("Fragmente," GA, 17: 201)

In telling us to turn to plants, Goethe emphasizes that the key to learning about gender lies not in separation of individuals or organs but somehow in their unity and their relationship to one another. Sexual parts of an organism are to be considered individual entities within one community. Masculine and feminine, male and female, cannot be understood if examined separately and apart from the context of the community of the whole organism. Plants suit the purpose because they often contain both sexes in one individual. In other words, if we automatically begin by presupposing differences, we will miss something crucial about what gender is and how it functions. By focusing on individuals like ourselves, who often appear to exhibit only one sex or gender, we concentrate too much on body parts and differences. Plants give us the opportunity to study both the similarities as well as the differences of the polar pair of the masculine and the feminine. They also allow us to track the immense competitive drive of each force.

Many of Goethe's contemporaries focused upon the complementary aspects of gender. For them, the feminine often represented the passionate and nurturing side of human society, whereas the masculine represented the rational and the active. Together, the two forces formed a complementary whole. The two sides, however, were not typically equal. Rousseau's *Emile* taught that women needed to be raised "specially to please man" and to be "subjugated" by him (358). Ultimately, women's function was to civilize men. Women were to so endear themselves to men that the men would become good family members and hence good citizens. Goethe's treatment of gender within many of his botanical works differs significantly from this view. Although he includes a discussion of complementary qualities, he also stresses the equality and the similarity of the sexes. In addition, his descriptions of gender, while following some of the characteristics prominent during his time, also present feminine characteristics as more free than masculine ones and hence the source for human fulfillment.

Although Goethe most clearly outlines the qualities and characteristics of the masculine and the feminine within his late "Spiral Tendency" essay, he already included a discussion of these issues in his very early "Metamorphosis" essay. Many scholars have been puzzled by Goethe's insistence on the equality of the two sexual parts within the "Metamorphosis."[16] In part, I believe this confusion has arisen because of the expectation that Goethe would favor the masculine over the feminine. As I have already argued in chapters 1 and 2, however, Goethe stressed time and again that the ideal was a balanced reconciliation of both parts. He writes in "Metamorphosis" that, because both organs are highly specialized and created at the same phase of development, the relationship of their development (die Verwandtschaft ihrer Bildung [no. 69]) is *outwardly* greater than that of any other part of the plant, i.e., the two parts resemble each other in appearance. At this stage in "Metamorphosis," both organs have reached their highly intensified stage. They appear to be ready to rule in a joint hierarchy. Their reunion symbolizes the climax of the plant's regular progression and will result in the creation of the seed and the continuation of the cycle.

While Goethe emphasizes the similarities between the forms and their stages of development, he simultaneously, albeit subtly, also calls our attention to their differences. The sexual parts are equal in that they both are heightened, and they are complementary in that they both have opposing functions within the process of reproduction. He also notes their differences. Essentially, he states that if one castrates the male filament by cutting off its anther, it resembles the female style (no. 69).[17] Not only is the anther that which produces/releases the pollen, thus paralleling the male sexual organ in animals, but in many plants, the anther even physically resembles the penis.

Goethe makes a very similar statement about the similarities and the differences of the sexual organs in his "Comparative Anatomy" essay of 1796. He believes that his animal type will be a useful basis of comparison, not only for comparing animals, but also animals and human beings, the various races, and most especially the sexes. This last comparison would enable a "deeper insight into the secret of propagation" [tieferer Einsicht in das Geheimnis der Fortpflanzung] (FA, I, 24: 272). He sees the two sexual organs as parallel and as illustrative of his highest concept (höchster Begriff): nature can modify and change identical organs so that they seem (scheinen) to be not only different in shape (Gestalt) and function (Bestimmung), but also, in a certain sense (in gewissem Sinne), as opposites (Gegensatz). Male and female sexual organs are similar because they arise at the same stage of development. Goethe here implies that the reproductive organs of animals, like those of a plant, arise from the same organ type. The similarities between the male and female sex organs equally demonstrate nature's heightened drive to come into being. As in "Metamorphosis," he simultaneously stresses their similarities as well as their differences. Their identical state of origin, mutual complexity, and heightened development make them equals. Their differing physical functions (as well as their ultimate difference in form) make them opposites.

Unlike Aristotle's ordering of nature, no parts or individuals are inherently or naturally higher than others. In "Ernst Stiedenroth" (1824), Goethe canvases his disagreement with the ancients. He discusses the various parts of the human soul and of nature at large and stresses their mutually and interconnected worth:

... I have often written of the dissatisfaction I felt in my younger years with the doctrine of lower and higher soul forces. In the human spirit, as in the universe, nothing is higher or lower; everything has equal rights to a common center which manifests its hidden existence precisely through this harmonic relationship between every part and itself. The quarrels in antiquity, as well as in modern times, all spring from a division of what God created as one in His realm of nature. ("Ernst Stiedenroth," 12: 45)

[... denn schon früher habe ich an mancher Stelle den Unmut geäußert, den mir in jüngeren Jahren die Lehre von den *untern* und *obern* Seelenkräften erregte. In dem menschlichen Geiste so wie im Universum ist nichts oben noch unten, alles fordert gleiche Rechte an einen gemeinsamen Mittelpunkt, der sein geheimes Dasein eben durch das harmonische Verhältnis aller Teile zu ihm manifestiert. Alle Streitigkeiten der Ältern und Neuern bis zur neusten Zeit entspringen aus der Trennung dessen was Gott in seiner Natur vereint hervorgebracht.] (FA, 1, 24: 614–15, original emphasis)

Goethe stresses the desirability of a harmonic relationship between opposing forces. His scientific works, however, show how difficult striking a harmony between these two forces may be, as each one may potentially try to overpower the other instead of striving for a harmonic reconciliation. Where Goethe's "Metamorphosis" essay introduces the role of the masculine and the feminine poles within the plant and by analogy within creative/productive processes in general, it does not, however, offer an analysis of the qualities of the masculine and the feminine poles themselves. In "On the Spiral Tendency of Plants" ["Über die Spiraltendenz der Vegetation"],[18] we learn even more about the relationship between the masculine and feminine forces within one individual. Where his earlier "Metamorphosis" essay discusses gender in conjunction with sexual organs, his later essay separates gender from body parts. Moreover, the "Spiral Tendency" completes and expands his earlier discussion by specifically addressing the issues of balance, war, and reconciliation between the masculine and the feminine forces as each side struggles to come into existence and create.

Masculine and Feminine Tendencies

Within Goethe's novels, gender lines are not necessarily drawn against sexual ones, but may instead show a certain amount of flexibility. Mignon's gender in *Wilhelm Meister's Apprenticeship* remains ambiguous for a significant portion of the novel, and she, as many of the female characters within the novel, dresses in male clothing. In the very beginning of *Elective Affinities,* Charlotte and Eduard define gender categories that appear quite fixed and follow traditional lines: women are more emotional and superstitious, whereas men more rational and active. During the course of the novel, however, Eduard follows the "feminine" characteristics and Charlotte the more "masculine" ones. Because Goethe envisioned an ever-changing, fluid nature, his scientific categories, like his social ones, are subject to greater flux[19] than those of his contemporaries, such as Schiller, who primarily emphasize the complementarity of the sexes. Goethe's discussion of sexual relationships and gender is a part of his political view of nature. The masculine and feminine are often at odds with each other, and one often tries to dominate the other without regard for the good of the whole. He details these gender wars in greatest detail within his "Spiral Tendency" essay. There he describes in detail the qualities and characteristics of each gender as well as their relationship to each other.

According to this essay, the masculine and the feminine are no longer confined to the sexual organs, but are linked to two separate and opposing life forces of the plant. Both of these forces strive to exert their influence upon the plant, and their influence is apart from the formation of sexual organs. These two tendencies, or two vital systems (die beiden lebendigen Systeme), affect "all plant structure, all plant formation in accordance with the law of metamorphosis" (Mueller translation, 137) [aller Bau, jede Bildung der Pflanzen, nach dem Gesetze der Metamorphose] (FA, 1, 24: 794).

The interaction of these two forces echoes once again the process of the union, separation. and reunion of the "Metamorphosis" essay. Here, however, the whole process, and not simply its culmination, specifically addresses the relationship of the masculine and the feminine:

Let us return now to general aspects and recall what we asserted at the outset, that the vertical and spiral systems are closely bound together side by side in the living plant. When we see that the vertical system is definitely male and the spiral definitely female, we will be able to conceive of all vegetation as androgynous from the root up. In the course of the transformations of growth, the two systems are separated, in obvious contrast to one another, and take opposing courses, *to be reunited on a higher level.* (Mueller translation, 145, emphasis added)

[Kehren wir nun ins Allgemeinste zurück und erinnern an das was wir gleich anfangs aufstellten: das vertikal- so wie das spiralstrebende System sei in der lebendigen Pflanze aufs innigste verbunden; sehen wir nun hier jenes als entschieden männlich, dieses als entschieden weiblich sich erweisen: so können wir uns die ganze Vegetation von der Wurzel auf androgynisch ingeheim verbunden vorstellen, worauf denn, in Verfolg der Wandlungen des Wachstums die beiden Systeme sich im offenbaren Gegensatz auseinander sondern, und sich entschieden gegen einander überstellen, *um sich in einem höhern Sinne wieder zu vereinigen.*] (FA, 1, 24: 804–5, emphasis added)

No longer is Goethe limiting the masculine and feminine to the development of sexual organs. Rather, he describes certain kinds of behavior and qualities as "masculine" and others as "feminine." These two forces are different and make different contributions to the whole organism. All plants, no matter what organs they develop, contain in some measure both masculine and feminine traits.

In the opening pages of the essay, Goethe presents the general characteristics of the masculine and feminine life forces. The qualities or characteristics that he assigns to each force, are, of course, in part based upon artificially constructed gender categories that were influenced by the conventions of his day. However, while some aspects of these categories support many traditional suppositions, other aspects also serve radically to challenge traditional boundaries. He begins with the vertical, or what he calls the masculine, tendency in plants. He asks us to consider it a spiritual staff (ein geistiger Stab) that comprises both the upright skeleton as well as the growth principle of the plant as it strives from node to node toward heaven. It is "the durable, eventually solidifying and permanent part" [das Bestehende seiner Zeit Solideszierende, Verharrende] of any plant or tree (FA, 1, 24: 787). This force is that which brings about a succession or a continuity of the whole (eine Kontinuität des Ganzen ["Über die Spiraltendenz," 1831, FA, 1, 24: 778]).

Unlike the vertical tendency with its rigid qualities, the spiral, or what Goethe calls the feminine, tendency is developmental, reproductory, and

nourishing (das Fortbildende, Vermehrende, Nährende [FA, 1, 24: 787])
and is the basic law (Grundgesetz) of life (FA, 1, 24: 788). The masculine
grows straight; the feminine grows in twists and turns. The masculine
tendency reaches its conclusion (Abschluß) in columnar growth (FA, 1,
24: 796); the feminine climaxes in the conclusion of flowering (Abschluß
des Blütenstandes [FA, 1, 24: 797]). The feminine's time reference is also
quite different than the masculine's. Where the masculine represents
that which is enduring, the feminine is that which passes away or is tran-
sitory (Vorübergehende [FA, 1, 24: 787]). Both forces struggle to stamp
their particular characteristics upon the plant: a plant normally strives to
grow upward, but often its leaves and stems are arranged spirally.

Sexual organs play only a minor role in the drama between the masculine and the feminine. The two opposing tendencies are not defined by
sexual organs. Sexual organs do not determine their differences; rather,
sexual organs result from a deeper division in the plant. Although these
tendencies are polar opposites, they, like polarity and Steigerung or subject and object, are also nevertheless inextricably linked with each other:
"Neither of the two life systems can be imagined alone; they are ever and
eternally one; *but in complete equilibrium they produce the most perfect vegetation*"
(Mueller translation, 132, emphasis added) [Keins der beiden Systeme
kann allein gedacht werden; sie sind immer und ewig beisammen; *aber im
völligen Gleichgewicht bringen sie das Vollkommenste der Vegetation hervor*] (FA, 1,
24: 787, emphasis added).[20] The most perfect vegetation arises only if the
masculine and the feminine tendencies can unite and coexist as equals.
Within the "Metamorphosis of Plants" essay, the meeting between
equals is a sexual union that Goethe further endows with philosophical
significance. In this latter essay, the perfect balance between plants is not
isolated to sexual union, but may be seen throughout the structure and
growth patterns of the entire plant.

Goethe stresses that we can best understand these two tendencies if
we imagine them working together. He gives us two visual images to describe their relationship. He first tells us to envision the image of a twining plant growing on a living staff (Stange). He likens this relationship
to the twining plant convolvulus wrapping itself around a staff, but only
if we imagine the staff to be living as well. Such an image is a graphic example (ein sinnliches Beispiel) and an analogy (Gleichnis) that can

come to our aid (FA, 1, 24: 798). At the close of his essay, he then goes one step further (einen Schritt weiter) and turns to a more intensified image of the same relationship. Where he initially portrays the masculine as a staff or pole and the feminine as a creeping plant wrapped around it, he now asks us to imagine an elm tree with a twining plant growing around it. Nature itself, according to Goethe, has recommended this image of elm and vine, where the masculine is the grantor (das Gewährende) and the feminine is the needy one (das Bedürftige). This graphic example, a recurring image in several different traditions, emphasizes marital discord (as in "Amyntas") as well as marital harmony. First, the image is very reminiscent of the caduceus or Hermes's staff, where two snakes are wound around a physician's staff, a symbol often used in alchemy to represent a reconciliation of opposing forces.[21] Second, it recalls the tree in the Garden of Eden, where the snake is poised to poison forever the relationships between human beings and God and between man and woman. In Jung's interpretation of this symbol, like Goethe's, the snake represents the vegetative life force.[22] Third, Goethe's image recalls the old marriage topos of the vine and the elm, a topos that, as Demetz points out, "constantly suggests, to writers of many ages, an idea as well as an ideal of marriage" (529–30).[23] The marriage between the elm and vine is ideal, because once the elm provides a skeletal structure for the vine, the vine can then produce more grapes than before.[24]

Because the two gendered tendencies are so competitive and driven in their quest to exert their own forms and existence, an ideal and balanced relationship is difficult to achieve. Although Goethe provides examples of such balanced relationships (as in his "Metamorphosis" essay), the poem that began this chapter, "Amyntas," gives some indication of what may go wrong when these two forces meet and do battle. Within the "Spiral Tendency" essay, Goethe describes how each of the forces constantly vies for the upper hand and how one is just as likely to triumph over the other. He therefore demands that we investigate:

where one or the other [system] reigns, for soon one, without overpowering its opposite, is overpowered by it, or they place themselves in an equilibrium. Through these movements, the qualities of this inseparable pair must become all the more apparent. (My translation) [wo eins oder das andere [System] walte, da es denn bald, ohne seinen Gegensatz zu überwältigen, von ihm überwältigt wird oder sich ins Gleiche stellt, wodurch uns die Eigenschaften dieses unzertrennlichen Paares desto anschaulicher werden müssen.] (FA, 1, 24: 795)

Neither force naturally rules over the other, but their natural relationship appears to be one of war. At stake is each force's existence. Each tendency would like the plant to exhibit more of its qualities. Moreover, the masculine and feminine tendencies in some plants "cannot stand each other's society" [sich mit einer solchen Sozietät nicht wohl verträgt] (FA, 1, 24: 788–89), and the stem leaves (outgrowth of the masculine tendency) and the eyes (outgrowth of the feminine tendency) race to beat each other to the top of the plant. In some trees, particularly those that have been overnourished, the spiral tendency interrupts the successive and straight growth pattern of the masculine tendency and causes the tree to grow in the shape of a curved staff (FA, 1, 24: 788, 800).

As each force tries to dominate, the disruption of balance between the two forces often leads to the creation of unusual forms and characteristics. Goethe gives several examples of such battles and of plants where one tendency has gained the upper hand over the other:

Other monstrosities, later to be presented in greater detail, arise when that upward-striving force is thrown out of equilibrium with the spiral and outdistanced by the latter. In such cases the vertical construction is weakened, and in plants which produce either fibres or wood, it is thwarted and almost destroyed; on the other hand, the spiral system, upon which the embryos and buds depend, is accelerated; the branches are flattened; and the plant stem, which is lacking in wood, is distended and its interior destroyed. All during this process the spiral tendency makes its appearance, expressing itself in windings, crooks, and twists. By studying these examples, one will obtain a basic text from which to draw conclusions. (*Mueller translation, 132)

[Auch andere Monstrositäten, die wir zunächst umständlicher vorführen werden entstehen dadurch daß jenes aufrechtstrebende Leben mit dem spiralen aus dem Gleichgewicht kommt, von diesem überflügelt wird, wodurch die vertikale Konstruktion geschwächt und an der Pflanze es sei nun das fadenartige System oder das Holz hervorbringende in die Enge getrieben und gleichsam vernichtet wird indem das Spirale, von welchem Augen und Knospen abhängen beschleunigt, der Zweig des Baums abgeplattet und des Holzes ermangelnd, der Stengel der Pflanze aufgebläht und sein Inneres vernichtet wird, wobei denn immer die spirale Tendenz zum Vorschein kommt und sich im Winden und Krümmen und Schlingen darstellt. Nimmt man sich Beispiele vor Augen so hat man einen gründlichen Text zu Auslegungen.] (FA, 1, 24: 788)

While the most perfect vegetation may arise through the balance between the masculine and the feminine, Goethe does not rule out the utility or the beauty of plants where one tendency has triumphed over the other. Nature's creative powers extend especially to them. Some plants possess predominantly only one tendency, so that the battle for power is futile. Goethe points to common flax as an example of the predominant masculine tendency at its best.[25] The strength of the masculine tendency is demonstrated by the enduring qualities of its thread. Indeed, the more the stem is exposed to the elements, the more beautiful and strong its thread.

Goethe's description of twining plants with predominantly feminine characteristics is significant for his philosophy of gender. Although, at times, the terms he uses to characterize the feminine are quite traditional, he also links the feminine with what is highest in human beings because it has more freedom of movement than the more structurally based masculine. For Goethe, the little forks of such creeping plants as grapevines best illustrate feminine characteristics. The forklets resemble their masculine counterparts of twigs, but lack their solidarity. Because they are so full of sap and so pliable, they show a special irritability (Irritabilität). *Irritability* was a charged term in the natural sciences and was closely associated with a Newtonian view of the world. Irritability[26] was a term used by Haller as "a force inherent in a particular type of matter (animal muscle tissue) that operates automatically upon proper conditions of stimulation" (Roe, 33). This term was used to describe such reactive movements as involuntary motion or the contraction of muscles in limbs that had been severed from the body. As one would expect given Goethe's diametrically opposed view of nature, he uses the term here quite differently than in Haller's sense. Although Goethe recognizes that some plants need an exterior stimulus to move (äußere Reiz), others, such as the tendrils of a passion flower, are able to move on their own. In fact, Goethe's essay shows that he is much more interested in the self-moving capabilities of vinelike plants. This kind of plant is so lacking in a masculine skeletal structure that it "seeks outside of itself what it ought to, yet fails to, provide for itself" [sucht das außer sich, was sie sich selbst geben sollte, und nicht vermag] (FA, 1, 24: 798). Due to

this very "lack," however, the feminine spiral vessels also have an independence and freedom of movement not accorded to the masculine: "An
independent life is given to them, and the power to move individually,
on their own initiative, and to assume a given direction" (Mueller translation, 140) [Es wird ihnen ein Selbstleben zugeschrieben, die Kraft sich
an und für sich einzeln zu bewegen und eine gewisse Richtung
anzunehmen] (FA, 1, 24: 799). Unlike the rigid, enduring masculine tendency as epitomized by the flax, the feminine, represented by the vine,
moves about more freely.[27] For Goethe, the feminine tendency carries
with it the animating and more creative principle (FA, 1, 24: 789, 790).
Through this principle, the plant "completes its life course and finally attains completion and perfection" (Mueller translation, 131) [wodurch
die Pflanze ihren Lebensgang vollführt und zuletzt zum Abschluß und
Vollkommenheit gelangt] (FA, 1, 24: 786).[28] Like the irregular barnacle,
the feminine tendency is more free of structure and rules. It is therefore
more free to express its will and its desires.[29]

In another botanical essay, "The Purpose Set Forth" ("Die Absicht
Eingeleitet," 1817), Goethe links freedom of movement, a characteristic
of the feminine tendency, to that which makes human beings the most
developed animal, whereas the masculine qualities of stability and en-
durance contribute to the best in the botanical world: "plants attain
their final glory in the tree, enduring and rigid, while the animal does so
in the human being by achieving the highest degree of mobility and freedom" (12: 65) [so daß die Pflanze sich zuletzt im Baum dauernd und
starr, das Tier im Menschen zur höchsten Beweglichkeit und Freiheit
sich verherrlicht] (FA, 1, 24: 394). The qualities that raise human beings
above plants and other animals arise from the feminine qualities of an
organism. Freedom of movement represents a heightened coordination
of the physical and spiritual parts of an organism, which in turn enables
an organism to fulfill its desires better and exert its will. Goethe also
speculates that plants and animals develop from opposing influences—
plants through light and animals through darkness.[30]

In the "Spiral Tendency" essay, Goethe further explores this theory of
light and darkness influencing development, as the masculine develops
in the light, the feminine in the dark. He provides a detailed account of
this gender difference of light and darkness when he describes the water
plant, Vallisneria, and its dual development. This plant proves to be an

interesting study because instead of blooming with one flower that has both male and female organs, this plant blossoms in singular flowers that are either male or female. Goethe turns to this plant as a "lucky example" and "highly meaningful" in the study of how the two systems develop side by side (FA, 1, 24: 802) The male individual (das männliche Individuum) shows itself on an upward-climbing straight shaft (FA, 1, 24: 802) that develops its organs only upon reaching air and light. The female individual spends much of its time under water and in the dark and ascends and descends into the water by tensing and relaxing its spiral stem. Because the plant is more developed and its male and female parts are divided, the "spiritual anastomosis" is even more tenuous than the one described in "Metamorphosis." The male individual emits thousands of fruit organs that spread out into the water like silver-white flakes and endeavor and strive (bemühen und bestreben) to reach the female. At the same time, the female emerges from the water and spreads open its crown. The flakes, in turn, strew forth their pollen and impregnate the female. Once impregnated, the female retreats to its underwater lair, where the seeds reach maturity. Once again, Goethe emphasizes the male as emerging out of light and air, while the female develops in darkness and moisture. Like the example of the "Mothers" in *Faust,* feminine creativity takes place away from the light of day, where it is allowed to mature and develop. The spiritual anastomosis combines not only the masculine and the feminine but also the polar opposites of light and darkness, air and water.

Goethe repeats this scenario of fertilization twice in order to be rid of the fairy tale (Märchen) that the whole male flower detaches itself and "lustily searches for a wife" [lüstern auf die Freite zu gehen] (FA, 1, 24: 804). In other words, the fairy tale holds that the male flower has freedom of movement. The female has the ability to move up and down, open and close. In addition, that the entire male individual does not detach to join with the female, but only its pollen, again deemphasizes the physical aspect of impregnation. As in "Metamorphose," this union is spiritual because the male never even comes into contact with the feminine. It emits only its "influences."

This chapter has focused upon nature's various means of coming into being. Organic parts as well as whole entities intensely strive to exist. While Goethe warns that the predominance of one individual does not necessarily lead to destruction or to negative results, he underscores the significance of a balanced reconciliation of the warring parties. In his poem "The Metamorphosis of Plants," we can further trace the significance of these competing desires for human beings and their endeavors.

Plants and Human Beings

In 1798, Goethe wrote a didactic poem, "The Metamorphosis of Plants" ("Die Metamorphose der Pflanzen"). In a way, it serves as an antidote to the extremely destructive relationship of "Amyntas": not only are the themes of both poems similar (the relationship between unequal lovers is likened to botanical images), but in his 1800 publication of these poems,[31] "The Metamorphosis of Plants" follows immediately after "Amyntas." "Amyntas" illustrates how destructive an imbalance of power may be, even when both partners are enjoying the imbalance. Although the masculine tree rejects the very possibility of ending the union with the feminine vine, the vine's mastery over the tree spells ruin for both: soon the tree will die and the vine will therefore have killed its own nutritive source. The "Metamorphosis" poem also starts with an imbalance of power. However, where the couple in "Amyntas" is doomed, the "Metamorphosis" poem describes how an equilibrium may be gained, so that the two disparate and willful entities may unite. Goethe's "Metamorphosis" poem, however, goes beyond the discussion of the relationships between the two competing forces of a plant and its analogy to two lovers. The poem also simultaneously discusses the reconciliation between the competing forces of science and poetry.

Goethe places his discussion of the poem within a scientific treatise, "Fate of the Printed Work." He begins by describing the disagreements that arose after he had published his essay on the metamorphosis of plants. Like the competing forces of a plant community, his theory became part of the community of competing thoughts:

However, once [an individual] comes forward with his opinion he will soon find out that various points of view compete with one another in the world, to the confusion of scholars and ignorant men alike. The times are split by factions that understand themselves as little as they understand their diametrical opposites. (Mueller, 170)

[Tritt er aber mit seiner Meinung hervor, so bemerkt er bald daß verschiedene Vorstellungsarten sich in der Welt bekämpften und so gut den Gelehrten als Ungelehrten verwirren. Der Tag ist immer in Parteien geteilt, die sich selbst so wenig kennen als ihre Antipoden.] (FA, 1, 24: 418)

Goethe laments the poor reception of his treatise. Scientists dismissed his work as poetry, while the women who appreciated his poetry, "were far from satisfied with abstract gardening" (Mueller, 172) [waren auch mit meiner abstrakten Gärtnerei keineswegs zufrieden] (FA, 1, 24: 420). Neither side wanted to admit "that science and poetry can be united" (Wissenschaft und Poesie vereinbar seien), and both sides had forgotten "that science had developed from poetry and they failed to take into consideration that a swing of the pendulum might beneficently reunite the two, at a higher level and to mutual advantage" (Mueller, 172) [Man vergaß daß Wissenschaft sich aus Poesie entwickelt habe, man bedachte nicht daß, nach einem Umschwung von Zeiten, beide sich wieder freundlich, zu beiderseitigem Vorteil, auf höherer Stelle, gar wohl wieder begegnen könnten] (FA, 1, 24: 420). In advocating a union between poetry and science, Goethe is not here abandoning his scientific stance. Rather, his incorporation of poetry emphasizes the inclusion of qualitative studies of nature alongside more quantitative ones. Gaining knowledge of nature once again mimics its processes. Goethe here, as in the "Spiral Tendency" and "Metamorphosis" essays, refers to a higher union between two opposing forces—a union that, when equal, ultimately benefits both sides. The two opposing sides (science and poetry, men and women) meet at a "higher" level through intensification.

Goethe quite literally seeks to combine poetry and science within this essay. Not only does the poem itself contain a scientific teaching within its story of two lovers, but Goethe inserts this elegy into a scientific treatise. The scientific poem, then, endeavors to reconcile readers of poetry to the science (FA, 1, 24: 420), while the inclusion of the poem within a scientific treatise attempts to reconcile scientists to poetry. This attempt to unite and reconcile poetry and science may be taken as one of the main characteristics of Goethe's scientific methodology. Scientific facts, regular laws, or mathematical formulas cannot fully account for nature's creative drives and spontaneous moments. That Goethe included poetry within his science, whether explicitly in this case or implicitly in many of his other scientific essays, reflects how seriously he took this proposition.

It illustrates as well why he had such difficulty in finding acceptance as a scientist.

The poem operates on at least three levels. Goethe claims to have written it to explain plant metamorphosis to women (FA, 1, 24: 420),[32] as a metaphor of his own relationship to his lover, Christiane (FA, 1, 24: 423), and to argue that poetry and science, like the divided and initially unequal lovers, may be united as equals. In other words, a plant's metamorphosis may symbolize not only the eventual balancing of two independent and conflicting wills, but also the conflicting perspectives of poetry and science. While Rousseau's educational goals are notorious for their subjugation of women to men, Goethe's educational goals (at least in theory) seem quite different. Instead of formulating an entirely different course of study for women and ultimately educating them to serve men (and thus in the end to serve political society as well), Goethe's educational plan within the poem aims at equality. He wants a full and equal reconciliation of the two sides, whether he is speaking of the union of himself with the beloved of the poem or the desired reconciliation of poetry (as represented in this essay by the female readers) and science (as represented by a more male-dominated sphere) within his scientific works.

The main part of the poem recounts the primary principles of the essay "The Metamorphosis of Plants." The movement of the poem begins with the intellectual separation of two lovers (or of poetry and science) and concludes with their equal union. The poem starts with the poet's beloved in a state of confusion (Verwirrung, the same word used above to characterize the state of scientific opinions). She is overcome by the numerous different kinds of flowers and is put off by their Linnaean nomenclature. Nature's forms seem infinite, and science has done nothing but to add to the confusion of forms. Like Goethe, who was fascinated and confused by the infinite shapes of clouds before discovering Howard's nomenclature, she does not know where to begin: "All the shapes are akin and none is quite like the other" (1: 155) [Alle Gestalten sind ähnlich, und keine gleichet der andern] (FA, 1, 24: 420). As we know from Goethe's meteorological works, the poet, as well as the scientist, try to "grasp" and "hold fast" nature's changing forms. The "Metamorphosis" poem similarly suggests that the same principle may satisfy both opposing groups. Underlying the confusion of forms is a secret law, a holy riddle (ein geheimes Gesetz, ein heiliges Rätsel) that perhaps will satisfy the conflicting demands of science and poetry and reconcile the poet's beloved to the study of nature.

The poet then summarizes the treatise, "The Metamorphosis of Plants." He outlines how the leaf changes into other forms, always becoming more and more articulated and specialized until the sexual organs are formed. Unlike the treatise, the poem often turns to anthropomorphic language to describe the process. The earth acts as a womb, the cotyledons represent childhood, and, most importantly, the sexual union of the plant's sexual parts is likened to a mass wedding:

> . . . the most delicate structures
>> Twofold venture [strive] forth, destined to meet and unite.
> Wedded now they stand, those delighted couples, together.
>> Round the high altar they form multiple, ordered arrays.
> Hymen, hovering, nears, and pungent perfumes, exquisite,
>> Fill with fragrance and life all the environing air.
> One by one now, though numberless, germs are impelled into swelling,
>> Sweetly wrapped in the womb, likewise swelling, of fruit.
>
> (1: 157)

> [. . . die zärtesten Formen,
>> Zwiefach streben sie vor, sich zu vereinen bestimmt.
> Traulich stehen sie nun, die holden Paare, beisammen,
>> Zahlreich ordnen sie sich um den geweihten Altar.
> Hymen schwebet herbei und herrliche Düfte, gewaltig,
>> Strömen süßen Geruch, Alles belebend, umher.
> Nun vereinzelt schwellen sogleich unzählige Keime,
>> Hold in den Mutterschoß schwellender Früchte gehüllt.]
>
> (FA, I, 24: 422)

This wedding of numerous couples highlights the union of the most developed, disparate individuals of the plant community: the masculine and the feminine organs of reproduction. They begin in separate striving, but the goal is the same for each. Nor is the wedding of the pairs a normal wedding with one couple, but is one where numerous couples appear together in front of the altar of the Greek/Roman god of marriage, Hymen, and where numerous offspring are created. The poet then explains the philosophical significance of the unions:

> Nature here closes the ring of the energies never-exhausted
>> Yet a new one at once links to the circle that's closed,
> That the chain may extend into the ages for ever,
>> And the whole may be infused amply with life, like the part.
> (1: 157)

> [Und hier schließt die Natur den Ring der ewigen Kräfte;
>> Doch ein neuer sogleich fasset den vorigen an,
> Daß die Kette sich fort durch alle Zeiten verlänge,
>> Und das Ganze belebt, so wie das Einzelne, sei.]
> (FA, 1, 24: 422)

Although each individual passes away, the union of the male and female results in the continuation of nature's creative processes. Each part of nature, like nature as a whole, is a continuous process of coming-into-being and transformation. Nature is characterized throughout by its creative energies. The individual masculine and feminine parts, the whole plant, and the species in general strive in their own ways to maintain their existence through reproduction. Goethe's attempts to poeticize one of his scientific treatises illustrates that, at heart, both poetry and science are concerned with these creative and dynamic natural activities.

As the poem continues, the poet and his beloved are also part of this same creative process. The poem concludes by addressing the role of human beings within the process of creative intensification. The ideal relationship between two people is like the progression of a regular plant:

> Oh, and consider then how in us from the germ of acquaintance
>> Stage by stage there grew, dear to us, habit's long grace,
> Friendship from deep within us burst out of its wrapping,
>> And how Amor at last blessed it with blossom and fruit.
> Think how variously Nature, the quietly forming, unfolding,
>> Lent to our feelings now this, now that so different mode!
> Also rejoice in this day. Because love, our holiest blessing
>> Looks for consummate fruit, marriage of minds, in the end,
> One perception of things, that together, concerted in seeing,
>> Both to the higher world, truly conjoined, find their way.
> (1: 159)

[O! gedenke denn auch, wie aus dem Keim der Bekanntschaft
    Nach und nach in uns holde Gewohnheit entsproß,
Freundschaft sich mit Macht in unserm Innern enthüllte,
    Und wie Amor zuletzt Blüten und Früchte gezeugt.
Denke, wie mannigfach bald die, bald jene Gestalten,
    Still entfaltend, Natur unsern Gefühlen geliehn!
Freue dich auch des heutigen Tags! Die heilige Liebe
    Strebt zu der höchsten Frucht gleicher Gesinnungen auf,
Gleicher Ansicht der Dinge damit in harmonischem Anschaun
    Sich verbinde das Paar, finde die höhere Welt.]
    (FA, I, 24: 422–23)

The relationship with the beloved begins with acquaintance, progresses to friendship, and then intensifies into love. Although their desires are initially separate, they become more and more united as their relationship progresses until they are finally in complete unison. The highest fruit of their union is not their child,[33] but a harmony of their minds.[34] Whereas a child symbolizes one aspect of nature's creative powers and represents the physical continuation of the chain of being, the poet expresses a desire to participate in an additional kind of union. He wishes to unite with his beloved on another level so that together they may explore a higher realm of existence. Their striving is to encompass physical as well as spiritual creation: creation of a child and of a unified perspective. While one may question the degree to which Goethe actually treated Christiane as an equal, in theory he stresses the equality of the two lovers. In his discussion of the poem, he remarks of its success in helping him achieve a better relationship with her: "And I too felt very happy to see that the living analogy intensified and completed our perfect attachment" [ich fühlte mich sehr glücklich als das lebendige Gleichnis unsere schöne vollkommene Neigung steigerte und vollendete] (FA, I, 24: 423). The poet claims success in that he has intensified his relationship in a way that is analogous to the manner in which a plant intensifies its organ of the leaf.

Although the poem is successful in uniting the two lovers and placing them upon an equal footing that allows them to find the higher world (höhere Welt), Goethe does not note equal success in uniting poetry and science in a higher place (höhere Stelle). Instead, he complains about how much his science has been misunderstood, so that "nothing is more painful than to have the things that should unite us with informed and intelligent men give rise instead to unbridgeable separation (Mueller, 174) [peinlicher kann uns nichts begegnen als wenn das was uns mit unterrichteten, einsichtigen Männern verbinden sollte Anlaß gibt einer nicht zu vermittelnden Trennung] (FA, 1, 24: 423). What he believed would profit from a reunion remained isolated. Goethe's hope was that his science, the science of a person known primarily as a poet, would help bridge the gap between poetry and science.

In this chapter, I have argued that Goethe's overarching theory of creativity focuses upon the various wills found within a pair, such as between two people, the elm and the vine, or poetry and the sciences. What appears to be whole and united will prove to be comprised of conflicting parts. Goethe's scientific conception of gender requires of a scientific observer to remain open to the unions, separations, and reunions of nature's parts. Nature does not remain static or necessarily follow regular patterns of development even within its gender and sexual categories. The parts of a plant are so free to exert their will that, during the lifetime of one plant, at times the masculine may predominate, while at other times the feminine gains power. Goethe's gender theory at times emphasizes the complementary relationship between the two genders (not unlike the more traditional feminist care ethicists, including Noddings, Baier, or Gilligan).[35] At other times, however, because he emphasizes the impossibility of purely objective science and argues for the necessity of constantly reformulating categories, his gender categories appear quite fluid (and bear some similarities with feminists such as Butler).[36] Nor does he limit himself to two genders. He also focuses much of his discussion on androgynous beings.

Many recent scholars of the eighteenth century have recently focused upon the link between the developing scientific interest in sexuality and the cultural and political changes of that era (Farley, 3; Pinto-Correia, 242–73; Schiebinger, 12, 38–41). Although this topic has never been extensively explored in the case of Goethe, I have shown that his discussions of human beings and the relationship between art and science reflect in many ways his treatment of the gendered forces within his plant studies. Although the parts of organisms necessarily struggle against one another and although an imbalance of power may even lead to the creation of new forms and entities, Goethe ultimately stresses that ideal conditions—whether in plants, in human relationships, or in theoretical perspectives—require the interaction of two opposing forces. The masculine and feminine tendencies of a plant, Faust's divided soul, two lovers, or even Goethe's own artistic creations—all presuppose divided wills whose very division is ultimately responsible for a striving to create.

CONCLUSION

GOETHE'S NATURAL PHILOSOPHY places nature's creative powers at its center. Each of his four main principles—polarity, Steigerung, compensation, and competition—focuses on nature's creative drives, whether on the small or the large scale.

In many of his works, these principles also explain important aspects of human creativity. Polarities thus equally explain a plant's progression, the formation of colors, the structure of Shakespearean tragedies, and certain aesthetic rules. Steigerung, once coupled with polarity, brings about more complex forms and ideas. In a plant, it explains how each individual plant strives to transcend its material base. For human beings, this natural principle provides the dynamic model for scientific and philosophical inquiry as well as certain literary endeavors. A novella thus may progress according to the same steps as a flowering plant.

Goethe's principle of compensation, in turn, accounts for even greater change and complexity than either polarity or Steigerung alone. According to this principle, the creative wills of natural beings struggle against the limits imposed upon them from without. These beings use the principles of polarity and Steigerung, in a give and take manner, to fight against these limitations and create new forms and ways of life for themselves. That new organic forms constantly arise illustrates the power of this creative will for Goethe. Similarly, Goethe's works describe how human beings use this same principle, whether in creating a beautiful work of art or attempting to achieve a psychological balance in their lives. Finally, the principle of competition demonstrates the fierceness of nature's will to create. In Goethe's conception of nature, each part of a larger whole is a potential revolutionary that may attempt to overthrow the established order. Nowhere is this fierce struggle more evident than in the masculine and feminine tendencies of a plant, which constantly battle each other for prominence. Goethe also paints a parallel image of this battle in his accounts of the struggle between the opposing forces of science and art and the conflicts between men and women.

Throughout this book, I have attempted to demonstrate that these four main principles set Goethe apart from both the Classical and Enlightenment traditions. He did not equate nature with necessity (or conversely human reason with freedom), because nature may be a free and creative entity (whereas reason may enslave). He questioned any assumption about nature or any scientific methodology that excluded nature's ability to change. Accordingly, he rejected teleology and polar hierarchies (masculine over feminine, soul over body, etc.) because these approaches, whether applied to natural or humanistic studies, were premised upon a static view of nature. For example, he rejected teleological approaches because nature may spontaneously create new forms and evolve into new beings. Although he admitted teleology's past political usefulness, he argued that the time had come to use models that would account for and embrace innovation and change. Similarly, he questioned studying nature according to causes and effects. He argued that to reduce nature to this formulation is to miss what is more interesting about it: how it creates new forms, how it overcomes obstacles, and how it strives to flourish.

Goethe was aware that scientific trends or models influenced many other spheres. Newtonian and Cartesian approaches, just as the Greek notion of nature centuries before, had a direct impact on politics, philosophy, and aesthetics. Teleology had served as the basis for hierarchical forms of government, whereas Cartesian duality had succeeded in separating and empowering reason and the mind above passion and the body, logic above creativity, and subject over the object. Goethe's scientific approach attempted to reunite what Descartes had separated. In advocating a reunion of these opposites, Goethe, in effect, was not only challenging the science and philosophy of his day, but the older Greek tradition as well. Although he praised the Greeks for their attempts to observe nature, he took strong issue with their ultimate characterizations of it. Plato, according to Goethe, had so emphasized the individual's interior, spiritual life that he failed to recognize the close ties with an ever-changing nature, whereas Aristotle had created a schema of nature that was so neatly ordered that he was incapable of seeing nature's creative potential.

Goethe, however, spent the greater part of his philosophic energies battling the Enlightenment tradition. To him, Newtonian and Cartesian scientists dealt with only one aspect of nature—its regular, physical laws. In his mind, their methodology had denied the validity of exploring any aspect of nature outside of mathematical models, and their approach had so dominated the discussion that it had become impossible for scientists to study nature in any other way.[1] As a consequence, Goethe argued that modern science impaired people from understanding (and sometimes even precluded them from seeing) color, creative drives, spontaneous impulses, evolution of form, competitive drives, and beauty. Because it was precisely these aspects of nature that he viewed as central to any understanding of it, he set out within his scientific texts both to criticize what he perceived to be the monolithic influence of Enlightenment science and to provide a new model for reincorporating nature's qualitative and dynamic aspects into scientific research. Instead of dissecting animals to explore nature's mechanics or analyzing discrete inorganic parts, his works attempt to examine whole organisms or complete phenomena in situ, i.e., within the context of their environment, their metamorphosing forms, and their creative activities. For him, dissection and analysis primarily show how individual organs, parts, or systems operate, whereas studying the entire natural being within its environment and its community of parts opens up a much broader field: one sees how organisms change over time, the strength of individual organic wills, and the general dynamic patterns of growth and development within nature.

The predominance of reason in both science and philosophy created, in Goethe's mind, stunted or misshapen personalities. His ideal human being was not a highly rational one, but a creative one, who would be able to balance and reconcile reason against passion and vice versa (FA, I, 24: 614–15). The Captain in Goethe's most scientific novel, *Elective Affinities,* is a caricature of the one-sided, rational man. Goethe makes fun of this engineer whose "extreme passion" is marked by the fact that he forgets to wind his precise watch. Faust, in contrast, represents the heightened desire to experience both life and rational knowledge. He is dissatisfied with the knowledge he has gained in the laboratory, and his quest becomes the attempt to find a balance between his two souls in the outside world. Similarly, it is not enough for the main character in the *Wilhelm Meister* novels to have the technical skill and training to be a good doctor. He must first learn about himself and about others before he can embark on, or even choose, his career of healing people.

Goethe also argued that the trust in regular laws, whether natural or social, gave people a false sense of security about their world and deadened their observational skills. Social structures, as the French Revolution demonstrated, could be disrupted at any time, just as a plant could deviate from its regular, progressive course. To trust in the permanence of institutions was to blind oneself to the warning signs of their decline or the tensions that could erupt at any moment. Similarly, to trust in regular, natural principles was to underestimate nature's power. The characters in *Elective Affinities* believe that they can control and tame nature: they graft shoots, they create new walks, and they recombine lakes. Their grafting, however, produces inferior fruit, their pathways make walking more difficult, and their creation of a new lake results in the drowning of one child and the near drowning of another. Because nature is not regular, it cannot be subdued by regular means. It is not, as one character wishes it to be, a static picture that may be viewed through a frame.

Goethe was, on the one hand, part of his time and a long scientific tradition. His scientific texts reflect and incorporate the ideas of other scientists and philosophers, from Aristotle's theory of compensation, to Schelling's and Hegel's interest in polarities, to Wolff's and Blumenbach's writings on the development of plants. Like Aristotle or Lucretius, Goethe bases many of his central tenets upon observations of nature. On the other hand, Goethe's scientific texts offer original ideas and penetrating criticisms. His natural philosophy, which adopts some aspects of ancient philosophy, adopts some aspects of his own time, and perhaps even anticipates still others of our own time, also contains criticisms of theories within all those times. The nature that he observes is not the hierarchical, ordered, or teleological entity of Aristotle. Instead, Goethe views nature as a willful, creative entity that may rewrite its own rules. However, while his notions of truth and hierarchy are not static (as were those of Aristotle or Plato), he (unlike many philosophers and theorists today) still maintains the existence of both. Because he viewed nature as forever recreating itself and transforming, he did not therefore believe that one can turn to nature to speak about final ends, absolute truths, or even future truths. One may only speak of present ones, and even those may be changing in front of our eyes.

Similarly, while Goethe's philosophy of polarity undermines certain categories, it is not therefore primarily a philosophy of negation, but one of coming-into-being. For Goethe, polarities are behind every creative act, from the production of color to the formation of plant organs. He also questioned the legitimacy of examining the world through polar hierarchies, whether those hierarchies support strict binary-based analyses (reason versus passion, masculine versus feminine, etc.) or notions of pure objectivity (scientific or otherwise). In our own time, many have done so largely by rejecting the use of nature as a standard of measure and questioning whether nature even exists apart from cultural constructs. Goethe similarly argues that the subject-object relationship is always influenced by the particular background and agenda of the subject. However, unlike many postmoderns, he believed that nature is not simply a cultural construct (although certain categories within it may be), but has its own independent and qualitative existence. Thus, natural objects also exercise a marked and at times even pathological influence upon the subject. Because of this interrelationship with nature, he also believed that we discover not only aspects of nature through scientific observation, but also ourselves. His scientific works, seeming at times to partake in anthropomorphism, often argue for a closely integrated view of nature, where human beings are influenced by and are creative through the same principles that operate within nature.

It is easy to see how Goethe's middle position has led to a great deal of misunderstanding in the analysis of both his scientific and literary works. His principle of Steigerung has led some to focus on his elitist tendencies, while those who turn to polarity are able to argue for democratic ones. If one examines his notion of the type apart from his analysis of evolving animal forms, one could argue, on the one hand, that he is a Platonic idealist, wedded to static forms. On the other hand, his characterization of nature's will closely resembles a Lucretian one, in which both organic and inorganic entities may at any given moment depart from a regular course. Goethe, however, differs from Lucretius in the important respect that Goethe does not view nature as simply spontaneous and random, but rather as possessing the ability to direct its will. Goethe's discussion of gender categories, analyzed apart from his polar principle, could very easily lead to the assumption that he followed Rousseau's lead on the dynamics between the masculine and the feminine. And, most importantly, in the history of the reception of his scientific works, his use of metaphysical language and symbols and his incorporation of poetry within his scientific texts have led many of Goethe's readers to dismiss his science as romantic—a science not based upon experimentation, but upon completely subjective feelings and emotions. However, a close study of Goethe's texts reveals that although he clearly departed from what is considered today as well as in his own day to be science, he did not do so by abandoning a commitment to scientific experimentation: he believed that mystical qualities could be reproduced by any researcher who was a "friend" of nature.

Goethe has long been dismissed as an unphilosophical or even an antiphilosophical thinker because he rejected philosophical systems and was critical of many philosophers. Helmholtz went so far as to say that Goethe's scientific works contain no theory whatsoever ("On Goethe's," 13–14) and that he was totally uninterested in understanding how nature works. For Helmholtz, Goethe's scientific endeavors were the activities of a poet who was "concerned solely with the 'beautiful show'" and not with what caused that show ("On Goethe's," 16–17). A study of Goethe's principles demonstrates that Helmholtz was wrong in believing that Goethe had no interest in researching the reasons behind natural phenomena. The main focus of Goethe's principles is to understand how nature works. Goethe's scientific works indeed demonstrate time and again that he rejects systems of any kind. These works also contain a dynamic philosophy that argues for the existence of creative and creating principles within nature. These principles do not represent closed systems, but an every expanding and changing world that constantly recreates itself and the rules that govern it. Nor were Goethe's views of nature, as many have argued, based primarily upon organic models (Gode-von Aesch, 154–55; Nisbet, 54, 134; Schmidt, 64). Rather, as his works on color and meteorology demonstrate, his dynamic principles applied equally to inorganic matter as to living forms.

Although Goethe may not have always practiced what he preached,[2] in theory he is extremely critical of those scientists who conduct science as "despots" (i.e., those who are dogmatic in their approaches). He argues instead that science should operate according to the principle of a "free republic" (FA, 1, 25: 32), where researchers are "free" to interpret data as they see fit (FA, 1, 25: 36). The concept of free republic is important to his scientific program because he thought that one could gain a more complete understanding of nature only by approaching it from multiple perspectives. Goethe argued that scientific results were often more telling about the psychology and the philosophy of the scientist conducting the experiments than about nature itself: a mechanistic philosophy, for example, presupposes a certain kind of scientific inquiry, and mechanists would tend to pose those questions for which their methods would be likely to find answers. Questions such as the cause of organic variations could be relegated to the "black box" and put aside.[3] In Goethe's mind, the biggest problem with a mechanistic view was that it represented just one perspective of the world, yet it was being treated as if it were the only way in which to approach the world. He therefore attempted through his scientific works to provide principles that emphasized an alternative scientific view: ones that examined different questions from those upon which mechanists focused. Goethe's questions, although often quite speculative, were aimed directly toward the black box. He attempted to discover principles that could account for nature's creative moments, whether in the creation of a regular flower or a new species. He focused his inquiries in those areas outside of a mechanistic program, such as an organism's striving to overcome obstacles and limits, the dynamic relationship of an organism's parts that could be understood only in the context of the whole organism, and an organism's active relationship to its environment.

For Goethe, nature's creative will was a scientific as well as a philosophic issue. A generation or so later, Nietzsche (admittedly in a different context) also thoroughly rejects a mechanistic (Darwinian) account of evolution and advances a view of life dependent upon the striving of a creative will:

But such a view misjudges the very essence of life; its will to power; it overlooks the intrinsic superiority of the spontaneous, aggressive, overreaching, reinterpreting and reestablishing forces, on whose action adaptation gradually supervenes. It denies, in the organism itself, the dominant role of the higher functions in which the vital will appears active and shaping. (*Genealogy of Morals,* part II, section 12:211)

Damit ist aber das Wesen des Lebens verkannt, sein Wille zur Macht; damit ist der principielle Vorrang übersehn, den die spontanen, angreifenden, übergreifenden, neu-auslegenden, neu-richtenden und gestaltenden Kräfte haben, auf deren Wirkung erst die "Anpassung" folgt; damit ist im Organismus selbst die herrschaftliche Rolle der höchsten Funktionäre abgeleugnet, in denen der Lebenswille aktiv und formgebend erscheint. (*Zur Genealogie der Moral,* 316)

NOTES

Preface

1. Several studies, for example, argue for the influence of earlier philosophical traditions upon Goethe's thought. Bell's book places Goethe's main literary and autobiographical works within the context of Enlightenment anthropology. However, Bell does not seek to find coherence in Goethe's thought on the "rigorously philosophical" level (2); instead, his study seeks to show that "Goethe's thought, viewed in historical terms, does cohere" (3). Prandi's study also seeks to place Goethe's ideas within the context of earlier thinkers. Her book argues more specifically for the influence of Lucretius and Spinoza upon Goethe's concepts of renunciation and happiness within his literary works. Similarly, Nisbet traces the earlier influences upon Goethe's scientific works in order to "assess Goethe's debt as a thinker to past traditions" (1). And although his study recognizes Goethe as a natural philosopher, as one who "refuses to divorce science from the rest of knowledge and experience," the primary focus of his book is to show the derivation of aspects of Goethe's thought from earlier Neoplatonic, empirical, and rationalistic traditions and not to analyze Goethe's own philosophy. Schmidt focuses his study on Goethe's pantheism, which he argues was much influenced by Spinoza. Gray extensively examines the influence of the alchemical tradition, whereas Schönherr traces the influences of the hermetic and Neoplatonic traditions generally, as well as the specific influences of Spinoza and Kant upon Goethe's color theory. Gode-von Aesch examines the background influences not only upon Goethe's scientific works, but also upon the works of Goethe's Romantic contemporaries.

2. Many earlier scholars have investigated the importance of Goethe's scientific projects for understanding his poetry and artistic thought processes. Although Wilkinson stresses Goethe's conviction that the "formative process is one throughout the universe" (148), she also argues that, in relation to philosophy, the "poet is much in evidence because it was to his experience of making poetry that the findings of any philosophy were referred" (147). Schaeder, in her very important and influential work, *Gott und Welt: Drei Kapitel Goethescher Weltanschauung*, argues that Goethe's nature studies are not an attempt to ground a natural philosophy, but an attempt to recognize the language of God within nature and to reconcile the two poles of God and the world (7–9; see also Wachsmuth, 90–91; 98). Like Wilkinson, she also stresses the importance of Goethe's natural studies for his understanding of art. Natural law becomes a key for Goethe with which to unlock the kingdom of art (e.g., 130, 153–60), and she therefore repeatedly turns to Goethe's scientific works in order to interpret his literary ones. See also J. Adler, Brodsky, Burwick (esp. 80–101), Huber, and Nygaard for additional examples of scholars who turn to Goethe's science to interpret specific literary texts.

3. Fink's seminal work examines how Goethe's understanding of boundaries and transitions informed his approach to the history of science. In the course of his book, he also analyzes, among other topics, Goethe's theory of biography, his narrative strategies, and the relationship between Goethe the scientist and Goethe the historian. The aim of Stephenson's work, which offers compelling linguistic analyses of several key scientific passages, is to identify "the cultural significance of Goethe's scientific writings" (vii). His book attempts to "rectify still persistent 'hyperbolic caricatures'" of Goethe's science (21) and examines Goethe's historical perspective on science. In Erpenbeck's overview of Goethe's science, he also places special emphasis upon Goethe's historical understanding ("Wissenschaft," 1193).

4. Several books have been written specifically about The Theory of Colors, from Sepper's significant study, which examines the specifics of Goethe's debate with Newton (Contra), to Schöne's, which treats The Theory of Colors as a literary, rather than a scientific or philosophical, work. Burwick's study focuses upon Goethe's optical writings and places them within the larger context of Romantic thought, broadly understood. Burwick thoroughly investigates the background influences on Goethe's optical works as well as the subsequent philosophical, aesthetic, and scientific reactions to them. See also Höpfner (Wissenschaft), who provides a detailed history of the reception of The Theory of Colors up through the twentieth century, and Gögelein, who focuses upon Goethe's optical works in order to analyze what Goethe means when he claims to practice science. During his investigation, Gögelein traces and examines the concepts behind Goethe's scientific methodology (such as Vorstellung, Urphänomen, and Erfahrung).

5. Steiner is considered by his followers to have "built a philosophical yet practical foundation for developing the latent possibilities of cognition explored by his predecessor [Goethe]." By using Goethe's precepts, "the student would be led toward an experience of 'the idea within reality'" (Riegner and Wilkes, 234). For example, Bortoft, after examining Goethe's understanding of the organic relationship of the parts to the whole, argues for the development of new educational systems that strive for a similar "authentic wholeness" and avoid the "counterfeit approach to wholeness" that Goethe was able to avoid. This style of "transpersonal education" would activate "a different mode of consciousness which is holistic and intuitive" and would lead students to "feel themselves to be more in harmony with the phenomenon" (24). Bortoft further suggests that knowledge of Goethe's science would bring about a "radical change in our awareness of the relationship between nature and ourselves. Instead of mastery over nature, the scientist's knowledge would become the synergy of humanity and nature" (115). This science of the "wholeness of nature" would further cultivate "a holistic mode of consciousness" (330).

6. Many other scholars have investigated (and questioned) Goethe's contributions to individual scientific fields. Early chaos theorists, such as Mitchell J. Feigenbaum and Albert Libchaber, were inspired by his nonmathematical, dy-

namic approaches of uncovering the secrets of nature. Feigenbaum, as a graduate student, read *The Theory of Colors* and "persuaded himself that Goethe had been right about color" (Gleick, 163–65), whereas Libchaber was interested in the "Metamorphosis of Plants" (Gleick, 197). See also Erpenbeck, who traces the influence of Goethe's scientific method upon several contemporary scientific theories ("Subjekt und Objekt," 221ff.). Neurologists now argue that Goethe's way of seeing color is a necessary complement to Newtonian rays of color (Sacks, 22–29), while botanists still debate the merits of his understanding of plant metamorphosis (Arber, 67–87; Coen and Carpenter, 1178–81; Flannery, 544–47; Lönnig, 574–76; Portmann, 133–45; see also Zirnstein). Edwin Land's color experiments, which led to the development of the Polaroid camera, bear a marked affinity to Goethe's color theories (Land, "Experiments," 84–89, and "Color," Burwick, 16, 51–52; 129; Sepper, *Contra* 14–16). Scholars have also investigated Goethe's contributions to anatomy (Brady, Haeckel, and Klumbies), geology (Wagenbreth), and (with a more critical eye) physics (Heisenberg and Helmholtz). Heisenberg even notes similarities between Goethe's conception of the archetypal plant and DNA: both are basic building blocks that attempt to explain the forces behind organic development ("Das Naturbild," 406–7). In the end, Heisenberg, however, does not believe that Goethe's science had a real influence upon the development of modern science ("Tradition in Science," 444).

7. See Sepper, who argues for the complexity behind such questions when investigating Goethe's and Newton's optical works (*Contra,* ix–x). He contends that much more is involved than "sorting out the details. The different fundamental aspects under which the question must be posed—phenomenal, theoretical, methodological, historical, and philosophical—must also be clarified" (ix). Sepper goes on to argue that one of Goethe's main points of contention against Newton was that Newton claimed that his theory was not an hypothesis, but indeed fact (20). Even Heisenberg, who contends that Newton was "right" in that he was "victorious" ("Goethe," 60, and "Die Goethe'sche," 146), maintains that "nothing can be gained from an investigation of their [Goethe's and Newton's] rights and wrongs" ("Goethe," 64, and "Die Goethe'sche," 150). Instead, Heisenberg focuses upon the complementarity of their approaches. Similarly, the quantum physicist, Walter Heitler, argues that Goethe's theories contribute to a qualitative understanding about color, a field in which mathematics has no place (66–71).

Introduction

1. Many have drawn comparisons between Goethe's theories and chaos theory (e.g., Gleick, 163ff.; Rowland, 93–107; Vazsonyi, 83–94).

2. Fink, who discusses Goethe's "persistent attention" to the topic of scientific language, argues that over time "a metalanguage of science began to emerge" (42). Stephenson also notes Goethe's complicated relationship to language and argues that Goethe "deployed a novel mode of writing . . . in order to re-enact, in lan-

guage at least, something of the complex interchange in natural process" (vii; see also 65–81).

3. Because of Goethe's emphasis on the interrelatedness of all of nature, many environmental ethicists see him as a precursor to their movement. Höpfner, for example, argues that Goethe's science is similar to "green science" (grüne Naturwissenschaft), which stresses an intact, integrated world. He states that Goethe's scientific method warns against the kind of science that enables the hole in the ozone layer, acid rain, the Chernobyl accident, etc. ("Wirkungen," 208ff.). Similarly, Altner argues that Goethe's scientific approach is "a true forerunner of alternative science" in that it focuses upon the close interaction and interrelation between human beings and nature (342). Muschg argues that Goethe's approach is ecologically based because it emphasizes a respect for the object (70–71). See also Schönherr, who discusses recent scholarship on the ecological aspects of Goethe's scientific method (8–9), and Heisenberg, who notes that "Goethe's fear" of modern science was justified in light of air, water, and soil pollution ("Tradition in Science," 444).

4. This list is by no means meant to be exhaustive. Other scholars have also argued for the importance of such concepts as "totality" (Totalität [Schönherr, 142; Teller, 129–33]), "ways of conceiving" (Vorstellungsarten [Gögelein, 77–82; Sepper, Contra, 91–99]), "archetypal phenomenon" (Urphänomen [Gögelein, 23–27; Schönherr, 106–11; Stephenson, 54–58; Teller, 132]), "experience" (Erfahrung [H. Adler, 272–74]), "cognition" (Erkenntnis [H. Adler, 277–80; Gögelein, 149–54]), and Goethe's propensity to use lists when describing scientific phenomena (Gögelein, 146–47; Pörksen, 138). However, not only are the four principles that I have selected the most prominent throughout Goethe's scientific corpus (and include at least tangentially the others already listed), but they also focus directly on nature's (as opposed to the scientist's) activities. For example, where compensation and competition focus specifically upon nature's actions and the development of forms, the archetypal phenomena act as a limit to scientific inquiry, whereas the ways of conceiving relate to the goal of scientific pluralism.

5. Wells, for instance, argues that it would be impossible for Goethe or any of his contemporaries to have been able to envision the principle of common ancestry (28). See also D. Kuhn (3–15), Wenzel ("Goethes Morphologie," 54–66, and "Goethe und Darwin," 1–5, 42–45, 66, 75, 92–94), and Magnus (116–18). Lenoir carefully distinguishes Goethe's approach to studying various animals from Darwin's ("External," 27).

6. Although Koerner discusses Goethe's botanical works, she bases her portrayal of his philosophy of gender primarily upon two poems and not upon his scientific works (493–95).

Chapter 1

1. Burwick discusses Goethe's and Schelling's mutual influence upon each other. Although they did not meet until 1798, Burwick argues that Schelling's *Von*

der Weltseele (1798) incorporated aspects of Goethe's early optical work, *Contributions to Optics* (1791–92). Burwick further notes that, after their meeting, Goethe was actively engaged with Schelling's work (126–27).

2. Engelhardt notes that "it is a well-known fact that the influence of Schelling's metaphysical natural philosophy dominates the Romantics' understanding of nature" (56). And while he notes that *Naturphilosophie* is not by any means homogeneous, Romantic scientists demanded that "physics and metaphysics should not be mutually exclusive, but interrelated" (56). See also Jardine, 231–33.

3. Literary theorists as well as historians of science tend to focus more upon this Goethean principle than upon any, whether discussing the structure of his literary works (e.g., J. Adler, Bennett, Huber, Lillyman, Milfull, and Wachsmuth), the status of his scientific thinking (Heisenberg, "Goethe" and "Die Goethe'sche"; Stephenson, 83–92; Helmholtz, "Goethe's Presentiments" and "On Goethe's"), the influence of Romantic science upon his scientific texts (Jahn, 76–77; Müller, 5; Schönherr, 115),or its relation to particular movements or figures, including Neoplatonism (Nisbet, 44–45), Hegel (Teller, 134), and alchemy (Gray, 75–100, 22ff.). Gögelein traces the etymology of the word and canvasses Goethe's use of it (105–9). Several scholars, including Göres and Wachsmuth, have analyzed the importance of polarities within Goethe's own life. For a general overview of Goethe's use of the term across his corpus, see Huber (863–65). Huber, however, does not view polarity as a force characterized by a desire to unite and create. He characterizes polar movements as "aimless" (planlos).

4. Bennett, in discussing *Faust,* explains that "the conceptual tool favored by Goethe for a basic scientific approach to the phenomenal world is 'polarity;' and the fabric of polarities (or as it were, material negations) by which concepts are distinguished and articulated" (70). Although Bennett discusses the importance of polarities for thought processes, he ultimately focuses upon polarities as destructive. He appears to treat Mephistopheles as a metaphor for polarity as a whole instead of just one part of it. Mephistopheles is related to the "negativity in the form of natural polarity" and therefore "represents not merely logical negation but the idea of nonbeing in general, and on the stage considered as world he represents not merely polar tension but a force tending toward universal destruction" (71). The subsequent striving that is produced by Mephistopheles' presence is an "utterly hopeless" one (73). Similarly, Miller argues that polarities in Goethe's novel, *Elective Affinities,* necessarily create a structure in which "[t]he novel's lines of self-interpretation contradict one another" (*Ariadne's,* 183), leaving the reader "without firm ground to stand on in interpreting the novel" (*Ariadne's,* 210). Miller therefore concludes that polarities (such as the opposition between Geist and Buchstabe or literal and figurative language) within the novel "interfere with one another" and that their "intertwining forbids a choice between one or the other, though they cannot be reconciled, put in a hierarchy, dialectically organized, or made the themes of a coherent narrative moving from one to the other" (*Ariadne's,*

221). Miller at times seems to grant Goethe consciousness of the contradictory aspect of his work, while, at other times, he implies that Goethe is unconscious of it (cf. "Buchstäbliches," 11, 13, 23; *Ariadne's*, 177, 183). Likewise, Burgard generally wishes to argue that Goethe consciously employed polarities as a means to undermine the surface meaning of his own text (60–65, 215–26). However, because these critics do not address polarity within Goethe's scientific works, they fail to address his emphasis upon polar creativity, i.e., that for Goethe polarities are not primarily negating forces but are at the source of all created acts.

5. In stark contrast to the American criticism, the more traditional German scholarship often focuses upon the polar subject-object relationship as evidence that Goethe completely trusted the senses. Such scholars see his highly descriptive science as a result of his emphasis on the interrelationship between the subject and object: scientists attempt to approach and understand the object by describing it as minutely as possible. According to this interpretation, Goethean scientists are not to question the background causes of the existence of a particular phenomenon or even to question its existence at all. For example, Helmholtz argued in the nineteenth century that Goethe had no scientific methodology other than listing, in steps, the phenomena that he observed, so that "thus we may attain an insight into their connection without ever having to trust anything but our senses" ("On Goethe's," 12). Goethe's attacks on Newton are to be seen in the light of a "forlorn hope, as a desperate attempt to rescue from the attacks of science the belief in the direct truth our sensations" ("On Goethe's," 14). In the twentieth century, Muschg writes that Goethe is certain that he may trust all of his senses, but especially his eyes (50). Truth emerges in the direct interaction between the subject's senses and the object (51). He further argues that the dignity (Würde) and respect (Ehrfurcht) characterizing the interaction between the two distinguish Goethe's scientific research from that of modern science (51). So, too, Schrimpf writes that for Goethe the subject-object relationship is not an abstract concept, but explains a human being's interaction with the world: the human telos is not to be discovered within oneself, but outside, in the world (128). In Schrimpf's account, the eye enables this interaction because it brings the whole object, as it were, to the subject (131). Similarly, Schmidt argues that Goethe's scientific thinking is imbued with "the pathos of the 'objectivity' of the truth as well as of world-reality" (41). Goethe's "objective-theoretical realism" is rooted in his trust of the senses (49–51). Heisenberg contrasts modern science's objective reality with Goethe's subjective one. Objective reality "proceeds according to definite laws and binding even when appearing accidental and without purpose," whereas in subjective reality "events are not counted but weighed, and past events not explained but interpreted" ("Goethe," 68, and "Die Goethe'sche," 153). Heisenberg further argues that Goethe bases his subjective reality upon the senses, especially the sense of sight ("Goethe," 63; "Die Goethe'sche," 149; "Das Naturbild," 396; and "Tradition in Science," 444).

6. Fink has already discussed this issue in relation to how scholars have falsely characterized Goethe's philosophy of history (149–51).

7. Wilkinson's discussion on the relationship between the subject and object focuses upon the issue of communication: "For Goethe, communication with the not-self, whether objects or persons, is the task to which man is called and which—as the author of *Werther* knew only too well—he relinquishes at his peril" (136).

8. Goethe did not title this essay, and it was not published in his lifetime. (It was probably written around 1799.) For a discussion of various titles given to it, see Fink (152).

9. The essay "Nature," upon which Goethe is here commenting, was for many years attributed to Goethe himself. It is now generally considered to have been written by the Swiss scholar and theologian Georg Christoph Tobler. The confusion over the authorship was Goethe's own: in his comments, given forty-five years after the essay itself was written, Goethe himself can no longer remember whether he had written it. For a discussion of authorship, see Boyle 1: 338–39.

10. Although Stephenson argues that Goethe was not original in using these terms because Newton was the first to coin the concepts of analysis and synthesis (5), one of Goethe's main targets in this essay is Newton and his scientific method.

11. Sepper, in a similar vein, argues that Goethe ultimately failed in presenting an "all-encompassing science of color" because he never "resolved the question of how mathematical conception and calculation are to be reconciled with seeing and experiencing the appearances" (*Contra*, x).

12. Of course, by extension, a purely (or a primarily) qualitative approach ought to be inadequate as well. Although it might be argued that Goethe approached his own science qualitatively to correct an imbalance, it is a serious flaw in his natural philosophy that he never himself presents a balanced scientific account that incorporates the two opposing methodologies. His own view, as his natural scientific writings show, remains qualitative. His statements about scientific balance remain on the theoretical plane.

13. The entire autobiography was written between 1809 and 1831. It is believed that Part II, the part containing Goethe's Neoplatonic view of creation, was written between 1809 and 1813 (HA, 9: 621). The creation myth occurs in Goethe's biography after he has recovered from his serious illness and while he is reading the works of several alchemists.

14. Mephistopheles similarly describes creation in *Faust I:* "I'm a part of the Part that first was all, / part of the Darkness that gave birth to Light" (2: 36) [Ich bin ein Teil des Teils, der Anfangs alles war, / Ein Teil der Finsternis, die sich das Licht gebar] (1349–50). Even Mephistopheles's explanation of his own existence matches the creation account: "A part of that force / which, always willing evil, always produces good" (2: 36) [Ein Teil von jener Kraft, / Die stets das Böse will und stets das Gute schafft] (1335–36). It is not that Mephistopheles does not *do* evil, because he clearly does throughout both parts of *Faust*. At the same time, his evil deeds are also directly linked with good ones—Gretchen's and Faust's redemption.

15. Moreover, this same pattern of contraction and expansion plays a prominent role in the development of the plant, and by analogy, the genesis of human
creativity (see chapter 2).

16. Schönherr argues that the relationship between light and darkness in this
poem is not primarily a polar relationship, but a demonstration of Goethe's notion of border or separation ("Idee der Grenze," 178).

17. For an explanation of Descartes's theory of matter, see Clarke, who explains, "The speculations about matter on which Descartes' theory of matter and,
subsequently, his concept of science depend include the assumption that the size,
shape and motion of small particles of matter would be adequate to explain all
their physical effects, including the physical effects on our sensory faculties which
stimulate sensations" (262).

18. This poem was part of Goethe's "Divan" ("Der West-östliche Divan") cycle
of poems.

19. Goethe similarly tells Eckermann (1 February 1827) that he does not repent
of the time spent upon his *Theory of Colors:* "I do not repent it at all . . . though I
have expended half a life upon it. Perhaps I might have written half a dozen
tragedies more; that is all, and people enough will come after me to do that" [Es
gereut mich auch keineswegs . . . obgleich ich die Mühe eines halben Lebens
hineingesteckt habe. Ich hätte vielleicht ein halb Dutzend Trauerspiele mehr
geschrieben, das ist alles, und dazu werden sich noch Leute genug nach mir
finden] (FA, 2, 12: 232). In another conversation on 2 May 1824, he compares his
work on colors to such previous revolutionary figures as Luther and Napoleon.
For similar comments, see Soret's notes on 30 December 1823 and Goethe's diary,
23 February 1831.

20. The most extreme view on Goethe's *Theory of Colors,* perhaps, is Kurt
Eissler's. He links Goethe's statement to the ego's narcissistic feelings often evidenced in dreams or to the attitude of delusional patients (1119). He further argues that Goethe's relationship to his own color experiments "has the characters
of paranoid psychosis" (1126). Scholars such as Finley simply assert that Goethe's
theory was not science at all (40).

21. Schönherr discusses in great detail the reasons behind Goethe's vehement
attacks on Newton's works. In particular, he focuses upon the importance of
Goethe's unified view of nature (esp. 134–40, 152ff.); see also Sepper, "Goethe
Against Newton" (175–93). Several authors argue for the validity of certain aspects of Goethe's theory. Sepper, for example, while granting Newton the "palm of
victory" for scientific achievement, argues throughout his book that Goethe "has
an ampler understanding of science than Newton" (*Contra,* x). For a thorough
comparison of Newton's and Goethe's methods, see Sepper's *Contra* and Ribe
(315–35). For a history of the reception of Goethe's works, see Höpfner
("Wirkungen," 203–11, and *Wissenschaft*) and Burwick, who extensively examines
the reactions of Goethe's British and German contemporaries (25–79), including
the reactions of artists (Runge and Turner), philosophers (Hegel and Schopen-

hauer), and scientists (Brewster, Purkinje, Seebeck, Steffens, and Young). Burwick also discusses how Goethe's theories were received by later scientists such as Helmholtz (64–65), Heisenberg (52), and Land (51–52). (See also Sepper's discussions of Helmholtz's and Heisenberg's interpretation of Goethe [*Contra*, 5–9].) Lauxtermann discusses Hegel's and Schopenhauer's reactions to Goethe's color theory (599–624), whereas Howells compares Goethe's, Schopenhauer's, and Chevreul's color theories. Erpenbeck ("Subjekt und Objekt," 22ff.) and Höpfner ("Wirkungen," 208ff.) argue for the influence on *The Theory of Colors* on a variety of scientific fields.

22. Nygaard, along similar lines, describes Goethe's assessment of the Newtonian world as "a world without taste, smell, color, or texture, that had been flattened out, robbed of essential qualities, and reduced to a two-dimensional graph, a set of formulas" (70). In her article, Nygaard's discussion of Newtonian science focuses on Goethe's use of signs and symbols: "What most perturbs Goethe about Newtonian science is that again and again, it allows the *symbol* to displace what it represents, the *sign* to displace the phenomenon as the real focus of attention" (70, original emphasis). See also Sepper, who emphasizes how Goethe was "combatting a defective conception of *science and scientific method*" ("Goethe against Newton," 182, original emphasis). Burwick writes that Goethe viewed Newton's attempts "to explain light and color in terms of mechanical laws" as "another instance of the invasion of mechanical philosophy upon the province of the human being" (24–25). Schöne takes a slightly different approach by arguing for the religious, rather than the philosophical, importance of Goethe's polemic against Newton (e.g., 20–21, 40–44, 52ff.). Pörksen takes issue with Schöne's characterization of *The Theory of Colors* as having a primarily theological bent. He describes Goethe's approach in that text as morphological (144).

23. Schmidt, in a similar context, turns to Heidegger to argue that the truth of a principle does not follow from its practical usefulness (15).

24. In discussing his theory of weather, Goethe uses similar language (FA, 1, 25: 275). Schöne links the language of color suffering (Leiden) to Christ's suffering (67).

25. For a detailed explanation of Goethe's refutation of Newton's famous prism experiment, see Sepper (*Contra*, 31ff.). Basically, Goethe argues that colors do not appear on a white wall except in that area in which the wall is bordered by a darker surface, i.e., that there needs to be a point of interaction of light and dark.

26. Paragraph numbers refer to numbers in the original and most subsequent editions.

27. For a detailed discussion of Goethe's attack on Newton's theories, see Sepper (*Contra*, esp. 81–84, 152–56), who goes through several of Goethe's arguments and experiments. See also Fink (31–36).

28. For an exemplary discussion of the issues involved, see Sepper (*Contra*, 46–59), who outlines Goethe's version of this experiment in his earlier optical work, *Beiträge zur Optik*. Sepper further argues that had Goethe been more open

about his intentions in his earlier work to "reopen the science of color to new lines of investigation and to reform the method of theorizing implicit in the Newtonian doctrine" (58), there might have been more appreciation in the scientific community for Goethe's emphasis on context and contrast (50–60).

29. This is an illustration from a woodcut that was modeled upon a drawing by Goethe. It was meant to be on the cover for the optical cards that accompanied his earlier work on optics, *Contributions to Optics* (1791). The drawing is of Goethe's own right eye, as seen in a mirror.

30. Stephenson, in analyzing a different passage, makes a similar point, very elegantly noting how the language of the passage mirrors the philosophical meaning of it. He remarks that Goethe's use of alliteration and repetition of words within the experiment illustrates "the Neo-Platonic notion that the organ of reception must . . . already possess the qualities it discerns in the world. . . . What cannot be rendered out of German is the sound-look play on the words used for 'surface' (Fläche) and 'spot' (Fleck), whose perceived virtual identity draws the whole web of subjects and objects together, making evident to the reader what the experiment makes manifest to the experimenter, namely that the 'coloured surface' ('farbige Fläche') and the '[white] spot' ('Fleck') are, in a meaningful way, the same, once perceived by the eye, whose property is to see the presence and absence of colour in objects, by dint of polarity" (77).

31. For similar comments on the reconciliation of opposites, see paragraphs 60, 334, 453, and 739.

32. In an essay on osteology, Goethe similarly writes that unless one learns to see with the eyes of the spirit (mit Augen des Geistes), one will remain blind in studies of nature (Naturforschung [FA, 1, 24: 248]).

33. Using similar language in an essay on meteorology, Goethe writes that every scientist strives to have a whole (zum Ganzen strebt [FA, 1, 25: 299]).

34. In "A Few Comments" ("Wenige Bemerkungen"), Goethe also discusses the two kinds of eyes. He is critical of Caspar Wolff's scientific method because it fails to realize that the spiritual eye must work in concert with the physical one: "[T]he worthy man [Wolff] nevertheless failed to realize that there is a difference between seeing and seeing; he failed to realize that the intellectual eye must work in constant and spirited harmony with the bodily eye, for otherwise the scholar might run the risk of looking and yet overlooking" (Mueller translation, 180) [so dachte der treffliche Mann [Wolff] doch nicht, daß es ein Unterschied sei zwischen Sehen und Sehen, daß die Geistes-Augen mit den Augen des Leibes in stetem lebendigen Bunde zu wirken haben, weil man sonst in Gefahr gerät zu sehen und doch vorbeizusehen] (FA, 1, 24: 432).

35. Schaeder similarly argues that totality for the eye is the same thing as harmony for the soul. In her discussion of the importance of the eye, while mentioning its ability to create, she stresses its independence (Selbständigkeit [266]).

36. In the "Metamorphosis of Plants," Goethe similarly endows the "eye" (Auge) of a plant with a kind of polar totality: if detached from the mother plant, it too will bring forth its own kind (nos. 85–93).

37. Goethe here seems to be echoing Aquinas's earlier interpretation of Aristotle's *De Anima,* which traces all thought processes back to the sense of touch.

38. In the *Phaedrus,* Socrates tells Phaedrus that he has no time to investigate, as scientists have tried to do, the truth behind the myth of Boreas. Socrates explains, "I myself have certainly no time for the business, and I'll tell you why, my friend. I can't as yet 'know myself,' as the inscription at Delphi enjoins, and so long as that ignorance remains it seems to me ridiculous to inquire into extraneous matters. Consequently I don't bother about such things, but accept the current beliefs about them, and direct my inquiries, as I have just said, rather to myself" (230a).

39. He further observes in another "Maxim," "Everything, that is in the subject, is in the object and a little more. Everything, that is in the object, is the subject and a little more" (my translation) [Alles, was im Subject ist, ist im Object und noch etwas mehr. Alles, was im Object ist, ist im Subject und noch etwas mehr] (FA, 1, 13: 219, nos. 2.83.1–2).

40. Stephenson makes a similar point, but emphasizes that Goethe's main point of contention with Newton is "a broadly cultural rather than a narrowly scientific one: Newton's manner of presentation will help displace our living perceptions of the world with the dead letter of abstractions" (26). Stephenson further argues that Goethe interpreted Newton's scientific method as a kind of psychological character flaw (27). After thoroughly comparing Newton's method with Goethe's, Sepper concludes that "Goethe's fundamental error may thus have been in a certain sense the complement of Newton's. If Newton reduced the study of color to a study of rays of light, Goethe tried to hold the study of light at arm's length from the study of color. Color science is in need of both, however" (*Contra,* 152).

41. See also Höpfner (208–11).

42. As Schrimpf notes, Goethe's understanding of the relationship between subject and object caused him to be overly critical of both Enlightenment science and the Romantic movement (143). Schmidt argues that Goethe rejected Kant and Fichte's philosophies as being too focused on the subject and not focused enough on the object (41–56). Heisenberg also argues that Goethe rejected Romantic art and literature for being too highly subjective and not rooted enough in the exterior world. Heisenberg further compares Goethe's rejection of Romanticism with his rejection of the abstract mathematical symbols of modern science ("Das Naturbild," 404).

Chapter 2

1. Goethe was quite familiar with Lucretius's *De Rerum Natura.* Excerpts from that work are included in the Historical Part of his *Theory of Colors,* and he also planned to write a poem modeled upon Lucretius's. He once even remarked in a letter to Friedrich Leopold Graf zu Stolberg that although he differs in some ways from Lucretius, he also adheres more or less to his doctrine (2 February 1789, FA, 2, 3: 457–58). Lucretius argues that since "nothing comes from nothing," human and animal will arises from a spontaneous swerve of an atom:

... at times quite uncertain and uncertain places, they [the first bodies] swerve a little from their course, just so much as you might call a change of motion. ...

... and if the first-beginnings do not make by swerving a beginning of motion such as to break the decrees of fate, that cause may not follow cause from infinity, whence comes this free will in living creatures all over the earth, whence I say is this will wrested from the fates by which we proceed whither pleasure leads each, swerving also our motions not at fixed times and fixed places, but just where our mind has taken us? (*De Rerum Natura*, lines 2.220–23, 253–60)

2. 5. For a discussion of Blumenbach's concept of *Bildungstrieb*, see Lenoir, *Strategy*, 20–24.

3. The German word "Versuch" has a myriad of meanings, including experiment, essay, and attempt. For a discussion of the double-edged meaning of the word in Goethe's work, see Burgard (106–11).

4. This latter aspect is discussed in greater detail in chapter 4.

5. Because Goethe's publisher, Göschen, refused to publish this essay, Goethe broke with him completely and turned instead to a publisher who was more supportive of the scientific works (Unseld, 71–73). As Goethe explained nearly thirty years later, "I found it hard to understand why he refused to print my brochure, when, merely by risking six sheets of paper at the very most, he might have retained for himself a prolific, reliable, easily satisfied author, who was just getting a fresh start" (Mueller, 169) [ich konnte schwer begreifen warum er mein Heft zu drucken ablehnte, da er, im schlimmsten Falle, durch ein so geringes Opfer von sechs Bogen Makulatur einen fruchtbaren, frisch wieder auftretenden, zuverlässigen, genügsamen Autor sich erhalten hätte] (FA, 1, 24: 416).

6. Botanists are still arguing Goethe's point (see Coen and Carpenter, 1178–81; Lönnig, 574–76). This theory was well known in scientific circles in the nineteenth century. Charles Darwin, for example, in his *Origin of the Species,* refers to it as "familiar to almost everyone" (cited in Cornell, "Goethe on Plants," 40–41).

7. As Overbeck has already noted, the term *metamorphosis* within the essay comes to represent both the cause and the effect of plant development. Goethe's essay tried to show not only *that* metamorphosis occurs, but simultaneously *how* it occurs: visual observations were turned into a theory about causes (46–47). Thus, where Heisenberg characterizes Goethe's *Theory of Color* as an effort to combine subjective and objective approaches ("Goethe" and "Die Goethe'sche"), Overbeck characterizes Goethe's "Metamorphosis" as the attempt to unite cause and effect (48). Similarly, Pörksen writes that Goethe tries even through his writing style to bridge the gap between idea (Idee) and experience (Erfahrung) (139).

8. Numbers after "Metamorphose" quotations refer to paragraph numbers of original text.

9. Although Goethe does not return within the essay to discuss the significance of white, he clearly sees white as evidence of advanced progression. His reference to white as representing a higher stage is somewhat puzzling. He does, however, discuss white within *The Theory of Colors.* At one point, he writes that white techni-

cally is not a color at all: "Everything living seeks color, individuality, specificity, effect, opacity even in its most finely divided parts. Everything which has died tends to white, abstraction, generality, clarity, transparency" (12: 253) [Alles Lebendige strebt zur Farbe, zum Besondern, zur Spezifikation, zum Effekt, zur Undurchsichtigkeit bis ins Unendlichfeine. Alles Abgelebte zieht sich nach dem Weißen, zur Abstraktion, zur Allgemeinheit, zur Verklärung, zur Durchsichtigkeit] (FA, I, 23, pt. I: 196, no. 586). Notably, the only other colors that he mentions in "Metamorphosis" are green and red (no. 103), the two colors that later in his *Theory of Colors* will come to represent true polar mixtures. These two colors also play a significant symbolic role at the end of *Faust,* when the angels strew red (purpur) and green roses (11, 707).

10. Goethe also uses the image of a ladder to describe the evolution of animal forms. In a conversation with Riemer (19 March 1807), the process of creating a complex animal seems analogous to the greater articulation of the leaf through several stages:

Die Natur kann zu allem, was sie machen will, nur in einer *Folge* gelangen. Sie macht keine Sprünge. Sie könnte zum Exempel kein Pferd machen, wenn nicht alle übrigen Tiere voraufgingen, auf denen sie wie auf einer *Leiter* bis zur Struktur des Pferdes heransteigt. So ist immer eines um alles, alles um eines willen da, weil ja eben das Eine auch das Alles ist. Die Natur, so mannigfaltig sie erscheint, ist doch immer ein Eines, eine Einheit, und so muß, wenn sie sich teilweise manifestiert, alles übrige diesem zur Grundlage dienen, dieses in dem übrigen Zusammenhang haben. (*Goethes Gespräche,* 2: 201, original emphasis)

Goethe emphasizes the creative capability of nature's parts. All existing beings have a historical past, and one must view them as having ascended from simple forms to more complex ones. Nature acts in a similar way, whether one is looking at simple or complex beings, because all parts of nature strive to flourish and create.

11. Schönherr, in sharp contrast, argues that Steigerung, whether in "Metamorphosis" or in *The Theory of Colors,* is inseparable from procreation. He does not see Steigerung as tending toward any goal but as a movement along a cyclical path (129). Schönherr has misunderstood the relationship between polarity and Steigerung. While both principles, according to Goethe's philosophy, are capable of procreation, Goethe distinguishes between the products of each. For instance, in "Metamorphosis," a plant may continue its existence in several different ways without undergoing intensification: if overwatered it will simply continue to grow without bearing fruit or it could be reproduced via stem cuttings. Reproduction, then, is not what is primarily at stake when Goethe describes the spiritual anastomosis that leads to the reunion of the sexes. For a discussion of the plant's two modes of reproduction, see Kirby (71–73).

12. Ironically, in Goethe's novel *Elective Affinities,* a child is born of two "parents" who never physically come together. The biological parents, Eduard and Charlotte, commit spiritual adultery while making love. Eduard holds "Ottilie" in his arms while making love to Charlotte, while Charlotte imagines a union with the

Captain while with her husband. When the baby is born, it resembles Ottilie and the Captain.

13. Numerous scholars have argued for a variety of influences upon Goethe's use of the image of the ladder for Steigerung. Gray extensively traces the influence of alchemical writings upon Goethe's use of this image in "Metamorphosis" (71–100). Schönherr points to the importance of hermetic writings, Spinoza, and especially Kant upon Goethe's principle of Steigerung (138–39). Fink argues for a more indirect influence of Kant: one which is mediated through Schiller (138ff.). I believe, however, that I am the first to link Diotima's ladder extensively to Goethe's philosophy. Several scholars have argued that Goethe was intimately familiar with the *Symposium,* but these scholars have primarily argued for the role of Aristophanes's speech in *Elective Affinities* (see Lillyman, 28–144; Milfull, 88). Friedrichsmeyer further argues that the character Diotima was familiar to "most literate Germans of the 1790's" and that she served as an idealized model for that time (122).

14. Gray also argues that Goethe's text has at least two different meanings. But where I argue that Goethe's texts are philosophical, i.e., that he bases his worldview upon his observations of nature, Gray argues that Goethe's views were predominantly based upon ideas gathered from alchemical, mystical writings (see also Portmann, 137; Schönherr, esp. 56–62, 100ff.). For example, Gray believes that Goethe's plant stages did not represent the stages as Goethe observed them, rather that he "repeatedly attempted to fit the life of the plant into a pre-conceived scheme" (97) based upon Boehme's system. Gray argues that Goethe was incapable of separating his scientific observations from mystical texts (97, 99). While I do not dispute the influence of several alchemical treatises upon Goethe's thought (Goethe himself describes their significance in his autobiography), Gray overemphasizes the role of alchemy in Goethe's scientific works. Gray must work around the problem of Goethe's repeated statements about the necessity of closely observing nature by claiming that such statements were "self-deception" (97). In addition, as we have already seen in "Experiment as Mediator" and *The Theory of Colors,* Goethe was extraordinarily conscious of the problems of trying objectively to study nature. In the end, Gray characterizes Goethe's scientific works as poetry (99).

15. In this poem, Goethe, as Linnaeus before him, draws parallels between the reproductive organs in plants and human beings. For a discussion of Linnaeus's language, see Farley (7–8, 118). In chapter 4, I return to this poem to discuss how Goethe takes the analogy one step further than his predecessor. The ideal union is represented as a union between lover and beloved. Its fruit is not simply children, but a higher meeting of the minds.

16. As in *The Theory of Colors,* Goethe uses eyes in a variety of ways to call our attention to the unity of nature. After he has discussed the role of botanical eyes in reproduction, he states that illustrations would be useful in order to place the theory before our eyes (no. 102). In another paragraph, he refers to our eyes and a plant's eyes in the same sentence (no. 103).

17. Stephenson further argues that Goethe's descriptive prose is closely linked to the processes being described. He observes that the language of Goethe's text in paragraph 42 reflects the close relationship between calyx and corolla: "In the German text, Goethe exploits certain untranslatable aspects of the language to produce the effect of a virtual identity between calyx and corona, and of an intertwining growth pattern between them. . . . The two-part relative clause spells out the two aspects of the calyx, but it is the climactic jingle 'Kelch, welcher' . . . which draws together into a knot the dominant sound- and look-play: 'Kelches-Nelke' (carnation)-'Kelche-welcher-Kelche-Kronen' (-blätter)—'Krone-Kelches', in the last half of which sound-chiasmus expresses the reciprocity of interaction in the now almost imperceptible difference between 'Krone' and 'Kelch'. We see, enacted in the very body of the (German) language, the ambiguous process in which calyces arise which could just as well count as coronae" (76).

18. In "Der Inhalt bevorwortet" ("The Content Prefaced"), Goethe also calls the products of scientific discussions "Früchte" (fruit) (FA 1, 24: 405).

19. An epigenesist "argued that each embryo is newly produced through gradual development from unorganized materials," while a preformationist "believed that the embryo preexists in some form in either the maternal egg or the male spermatozoon" (Roe, 1; see also Farley, 16–33; Foucault, 125–65; Roe, 45–88).

20. Some scholars, such as Koerner, downplay the importance of Goethe's interest in biology. She characterizes it as being "embedded in the social spaces of holiday outings, tourist trips, hunting parties, spa hotels, and public parks" (476) and "part of a burgeoning holiday industry for scientific tourists" (477). And even though Goethe published numerous essays on botany, Koerner maintains that "Goethe's botanic project remained a private hobby" (487). She further compares Goethe's search for the archetypal plant to the flower-picking madman in *Werther* (482): his search for this plant, as his entire science, is "solipsistic and anti-empirical" (484).

21. This meeting may also be accomplished with a simple prism experiment. If one views black strips against a white background through a prism, and one turns the prism so that the white spaces are filled with color, the colors will meet at the center to form red (nos. 209–17). Or conversely, one may look at the bars of a window through a prism and receive the same result (no. 216). This same experiment conducted with white strips upon a black background will yield green.

22. The three parts of the whole opus—the Didactic, the Polemic, and the Historical—may also be seen in light of a ladder. The Didactic Part is the pinnacle and most forward looking one as Goethe presents the experiments that he believes will change the way in which we think of color and optics, the Polemic in attacking Newton seeks to unseat the present authority in the field of physics, and the Historical looks to the past to establish what groundwork has already been laid. The Historical Part, also follows a line of progression as it begins with the ancients and concludes in the eighteenth century. Schöne further argues that the Historical Part is modeled upon Arnold's church history (55).

23. See also M. Stern, who argues that Goethe's *Theory of Colors* investigates the relation of science and literature (247).

24. Pörksen further argues that Goethe uses grammatical *Steigerung* within his *Theory of Colors* and other scientific works. He sees that Goethe's frequent use of comparative, semantic lists as evidence of philosophic *Steigerung*. For *The Theory of Colors,* he gives the example of "Anblicken—Ansehen—Betrachten—Sinnen—Verknüpfen—Theoretisieren" (138).

25. Both Fink (42—55) and Stephenson (65—81) discuss Goethe's philosophy of language. Both especially focus upon his analysis of the inadequacies of language in describing scientific phenomena. See also Pörksen, who gives several examples of Goethe's attempts within his scientific writings to circumnavigate the static nature of terms and concepts and make them more fluid.

26. Goethe, in a conversation with Eckermann, states that the divinity (Gottheit) reveals itself in both physical and moral "Urphänomene" (13 February 1829).

27. The word *Vermittlung* is an important one for Goethe's scientific method. In his essay, "Der Versuch als Vermittler von Objekt und Subjekt" ("Experiment as Mediator Between Object and Subject"), the experiment bridges the gap between subject and object, so that one can learn that "[nothing] happens in living nature that does not bear some relation to the whole. The empirical evidence may seem quite isolated, we may view our experiments as mere isolated facts, but this is not to say that they are, in fact, isolated. The question is: how can we find the connection between these phenomena, these events?" (12: 15) [In der lebendigen Natur geschieht nichts, was nicht in einer Verbindung mit dem Ganzen stehe, und wenn uns die Erfahrungen nur isoliert *erscheinen,* wenn wir die Versuche nur als isolierte Fakta anzusehen haben, so wird dadurch nicht gesagt, daß sie isoliert *seien,* es ist nur die Frage: wie finden wir die Verbindung dieser Phänomene, dieser Begebenheiten?] (FA, 1, 25: 33, original emphasis).

28. In a 1809 diagram ("Schema der Seelenkräfte"), in which Goethe matched the powers of the soul with colors, green is termed "useful" (nützlich) and is a result of the blending of sensuality (Sinnlichkeit—blue) and understanding (Verstand—yellow). In a later color tetrahedron (1816—17) that Goethe constructed to symbolize the powers of the soul, green is symbolic of sensuality (Sinnlichkeit). Although the symbolism changed slightly over the years, green is linked to procreation in either example (LA, 2, 4: 329—30).

29. Red, in the diagram that Goethe and Schiller's created of the soul, is termed "beautiful" (schön) and is the result of the union of fantasy (violet) and reason (orange). In Goethe's later color tetrahedron, red becomes identified with fantasy alone. In both cases, however, red is linked with poetic creation and imagination.

30. In chapter 4, I discuss in more detail how the creative/procreative striving of plants relates to Goethe's views on the masculine and the feminine.

31. In this respect, Goethe's natural philosophy, with its departure from an anthropocentric perspective, could be seen as a precursor to the modern environ-

mental ethics movement. Like the environmental ethicists, Goethe warns against
the consequences of placing human beings above or outside of nature. For a parallel discussion, see, for example, Noddings's discussion of caring for animals (*Caring*, 148ff.). Unlike some environmental ethicists, however, Goethe does not see all
of nature as equal. Although all natural entities have inherent worth, those that
have intensified are considered higher than those that have not.

32. The quote comes from an unpublished introduction to the first three parts
of *Poetry and Truth*: "Ehe ich diese nunmehr vorliegenden drei Bände zu schreiben
anfing, dachte ich sie nach jenen Gesetzen zu bilden, wovon uns die Metamorphose der Pflanzen belehrt. In dem ersten sollte das Kind nach allen Seiten zarte
Wurzeln treiben und nur wenig Keimblätter entwickeln. In zweiten der Knabe
mit lebhafterem Grün stufenweis mannigfaltiger gebildete Zweige treiben, und
dieser belebte Stengel sollte nun im dritten Band ähren- und rispenweis zur Blüte
hineilen und den hoffnungsvollen Jüngling darstellen" (FA, 1, 14: 971). In a conversation with Eckermann, Goethe uses very similar imagery to discuss the development of children (6 March 1831, FA, 2, 12: 456−57).

33. Goethe, of course, at times did indeed use teleological language, as his discussion on his "Novella" demonstrates. Goethe, however, is fairly consistent in his
scientific works in opposing a teleological approach to studying nature. For a more
strictly teleological interpretation of Goethe's biology, see Lenoir, who states that
"Goethe's biology is fundamentally and radically teleological in character" ("Eternal Laws," 18). Although Lenoir argues that Goethe's view of nature is more active
than Darwin's, he still contends that "Goethe's universe is based on the rational
relationship of ends to means. To argue that the principal source of change in organic nature ultimately depends on chance is, in Goethe's view, to surrender the
goal of achieving a scientific treatment of biological organization" (27). Lenoir,
however, does not describe Goethe's teleology as linear, but as reflexive, where
"cause and effect are so mutually interdependent that it is impossible to think of
one without the other" (19). In his discussion of Goethe's history of science, Fink
sees a connection between Goethe's use of teleology and his attempts to contextu-
alize his work. He argues that Goethe is following Schiller's philosophy of histori-
ography. Schiller, turning to Kant's formulation of teleology, argued that the "histories of mankind must be written according to the principles of teleology," but a
teleology "in which the determination of divine providence was removed, condi-
tioned instead by a concept of freedom" (Fink, 139). Fink goes on to argue that
Goethe hoped that his essays would shape the posterity that would receive them—
a practice that Goethe began with his scientific works, but then carried on in his
literary ones (141). Although I disagree with Lenoir's characterization of Goethe's
teleology, Lenoir's book (*Strategy*) provides an excellent background to the discussion of teleology generally during this period. Within the book, Lenoir discusses
several different interpretations of teleology that were current in Goethe's time.
He defines the classical version as one that views the universe as "fundamentally
biological. Not only is each part subservient to the organization of the whole, but

there are only 'biological' laws. . . . Each part [of the cosmos] had its proper position relative to the whole which determined its natural motion" (*Strategy*, 10). Lenoir traces this philosophy in the works of the Naturphilosophen and Hegel.

34. This point will become clearer in the next chapter, which investigates how Goethean organisms change and evolve over time.

35. "Die Vorstellungsart: daß ein lebendiges Wesen zu gewissen Zwecken nach außen hervorgebracht, und seine Gestalt durch eine absichtliche Urkraft dazu determiniert werde, hat uns in der philosophischen Betrachtung der natürlichen Dinge schon mehrere Jahrhunderte aufgehalten, und hält uns noch auf . . ." (FA I, 24: 210).

36. Nisbet, in his discussion on teleology, argues that Goethe was influenced by Leibniz, Bonnet, and Herder in his views (59). Nisbet, however, argues that while Goethe did reject an "external purpose" to which an organism may conform, "he did not reject teleology in so far as it is *internal* to the organism" (59, original emphasis). Goethe's views on evolution, as the next chapter argues, illustrate that organisms are able to break free of internal teleological forces.

37. In his essay on comparative anatomy, Goethe similarly discusses how teleology has held back scientific progress (FA, I, 24: 228).

38. Goethe's worldview resembles that of some modern environmentalists. See, for example, Altner, 341–50; Höpfner, "Wirkungen," 208ff.; Muschg, 70–71; Schönherr, 8–9.

39. Goethe's essay and Kant's *Third Critique* were both published in 1790.

40. For a discussion of Kant's teleology and its relation to physico-mechanical principles, see Lenoir (*Strategy*, 22–35). For a discussion of how Kant's teleology was often misinterpreted by English scientists, "who suppose that 'mechanism' is the 'real reality' beneath the phenomena, while teleology is a sort of illusory human projection," see Cornell ("Newton," 408–9).

41. Gleick argues that Goethe's work inspired several chaos theorists (163–65, 197). M. Stern argues more generally that Goethe's science is compatible with postclassical physics, especially quantum physics (247–50).

42. Several physicists have argued that nonphysicists have seriously misunderstood the ramifications of quantum physics: "the consequence can be an exaggerated despair over the possibility of reasoned inquiry, or, even worse, an outright celebration of unreason and incoherence" (Gross, 117). Rather than urging that quantum physics is a "validation of the postmodern viewpoint," physicists such as Sheldon Goldstein (120–25) and J. Bricmont (131–37) argue that one needs to readjust the way we interpret nature to include chaos and pattern together.

Chapter 3

1. For a more complete discussion of the characters' relationship to money, see Tantillo ("Deficit").

2. For an analysis of the fluidity of the archetypal organ of the leaf in this context, see Brady (269–87), who argues that an emphasis on the static function of

the leaf has been caused through a misreading of Goethe through the lenses of Oken's and Owen's biology.

3. Lenoir, too, rejects the notion that Goethe was an essentialist. He, however, approaches the problem somewhat differently by equating the type with morphological laws. He argues that Goethe's morphotypes are "necessarily hypothetical relations arrived at through what Goethe describes as 'der spekulative Geist.' But they are not for that reason less really present in nature. In Goethe's view it is imperative to note that nature operates in terms of *forces* and *laws*. Both are present in nature, but in different senses" ("Eternal" 23, original emphasis).

4. Barsanti, in contrast to Mayr, argues that Lamarck's research demonstrates a gradual shifting away from mechanistic philosophy. While arguing that Lamarck was clearly a materialist, Barsanti contends that Lamarck plotted a middle course between the animists and the mechanists (54)—a course that recognized "that the phenomenon of life cannot be reduced to its mere physical-chemical aspect" (59).

5. Engelhardt describes how the idea that the "mutation of species as a genealogical transformation to external influences" is generally attributed to Lamarck around 1800 (57). Mayr further explains Lamarck's view: "Since a species must be in complete harmony with its environment and, since the environment constantly changes, a species must likewise change forever in order to remain in harmonious balance with the environment" (349).

6. Mayr notes that Darwin's methodology was based upon Newton's in that Darwin looked "everywhere in natural phenomena for laws, and in particular . . . for mechanisms or causes that were able to explain phenomena in widely different areas" (436).

7. Appel writes that this debate "brought to a head . . . a fundamental division in the biological sciences: whether animal structure ought to be explained primarily by reference to function or by morphological laws" (2).

8. Appel explains that, for Cuvier, "every animal part . . . was designed to contribute to the functional integrity of the animal or to adapt it to its environment. Whenever a new part was required for functional purposes, the Creator was free to fashion a new and appropriate organ. For Cuvier, function was paramount: the animal's needs sufficed to determine its structure" (4). Geoffroy, in contrast, "believed one could discover a generalized vertebrate anatomy, a single structural plan that could be traced throughout the vertebrates. The key to determining the ideal plan . . . was to ignore the form and the function parts and concentrate instead on the connections between parts. . . . the unity of plan in the animal kingdom preceded particular modifications of the plan to suit functional requirement" (4).

9. Goethe also wrote two reviews of the debates (FA, 1, 24: 810-42).

10. Goethe formulated his theory before Geoffroy, but it is unlikely that Geoffroy knew of Goethe's works. Once Geoffroy learned of Goethe's support, however, he actively promoted Goethe's theories (Appel, 159; see also Mayr, 463, and Brady, 258-59).

11. Fink has already argued for the importance of nature's transitional moments for Goethe. In his book, he examines how these transitions defined Goethe's approach to the history of science (see esp. 12–25 and 24–41).

12. For a discussion of Goethe's theory of compensation within the scientific context of its time, see Wells (19–23) and Magnus (101ff.).

13. Goethe similarly hopes that his *Theory of Colors* would be of use to people in disparate fields, including doctors, painters, chemists, and technicians.

14. For an in-depth discussion of Goethe's work on the intermaxillary bone, see Bräuning-Oktavio, who thoroughly examines Goethe's sources and influences as well as his relationship to the other scientists of his day. See also Fink, who discusses Goethe's discovery and its relevance to his views of natural harmony (21–25). Müller-Sievers examines Oken's claim to the priority of discovery.

15. See, for example, Aristotle's *Parts of Animals*: 655a18–34, 657b5–35, 663a30, 664a1–14, 674a32–674b18; and his *Generation of Animals*: 749b5–750a5, 750a21–35. Graham also points to Plato's *Protagoras* as one of Goethe's sources in developing his theory of compensation (153–54). Stephenson links Goethe's idea to Leibniz (5).

16. Goethe, of course, is wrong in his supposition—some animals have both horns and tusks (see Cornell, "Faustian," 488).

17. Goethe uses almost identical language in his notes on plant development: "die Ausdehnung des einen Teils ist Ursache daß ein andrer Teil aufgehoben wird. Zum Grunde dieses Gesetzes liegt die Notwendigkeit an die jedes Geschöpf gebunden ist daß es nicht aus seinem Maße gehen kann. Ein Teil kann also nicht zunehmen ohne daß der andere abnimmt, ein Teil nicht völlig zur Herrschaft gelangen ohne daß der andere völlig aufgehoben wird" (FA, 1, 24: 103).

18. In this respect, Goethe differs from Darwin. Darwinian organisms vary spontaneously, whereas Goethean ones vary partly in reaction to their environment (see Lenoir, "Eternal," 27).

19. In his introduction *(Einleitung)*, d'Alton cites Goethe as the trailblazer in the field of metamorphosis. D'Alton here freely adapts Goethe's terms for plant metamorphosis to account for changes in animal forms. D'Alton does not use the term *Typus,* but does more generally focus upon the relationship (Verwandtschaft) of all animals to one another, both past and present (Einleitung, n.p.)

20. It is interesting to compare this passage with one in Nietzsche's *Genealogy of Morals* (Part II, Section 16). Using very similar language as Goethe, Nietzsche describes the evolution of water animals to land creatures. Nietzsche then uses the example of this adaptation as an analogy to explain the development of human consciousness. As human beings evolved from four-footed animals to two-footed ones, other adaptations followed: "Of a sudden they found all their instincts devalued, unhinged. They must walk on legs and carry themselves, where before the water had carried them. They felt inapt for the simplest manipulations, for in this new, unknown world they could no longer count on the guidance of their unconscious drives. They were forced to think, deduce, calculate, weigh cause and ef-

fect—unhappy people, reduced to their weakest, most fallible organ, their consciousness" (Golffing translation, 217) [mit Einem Male waren alle ihre Instinkte entwerthet und "ausgehängt." Sie sollten nunmehr auf den Füssen gehn und "sich selber tragen," wo sie bisher vom Wasser getragen wurden: eine entsetzliche Schwere lag auf ihnen. Zu den einfachsten Verrichtungen fühlten sie sich ungelenk, sie hatten für diese neue unbekannte Welt ihre alten Führer nicht mehr, die regulirenden unbewusst-sicherführenden Triebe,—sie waren auf Denken, Schliessen, Berechnen, Combiniren von Ursachen und Wirkungen reduzirt, diese Unglücklichen, auf ihr "Bewusstsein," auf ihr ärmlichstes und fehlgreifendstes Organ!] (322).

21. Today, it is generally agreed that the "forms of heritable variation that arise are not causally dependent on the nature of the world in which organisms find themselves" (Lewontin, 47–48). Lewontin stresses, however, that a more interdependent picture of an organism and its environment needs to be developed by modern biologists: "But the claim that the environment of an organism is causally independent of the organism, and that changes in the environment are autonomous and independent of changes in the species itself, is clearly wrong" (48).

22. Schiebinger notes that "few natural historians went as far as Linnaeus" in placing apes and human beings in the same order and that "most naturalists maintained that while apes might bear human characteristics, they certainly were not human" (81).

23. See also Cornell ("Faustian," 486ff.).

24. Schiebinger notes that upright posture has been considered a human trait since the time of Plato. She also describes the eighteenth-century debate over whether apes had upright posture. Although Buffon, for example, depicted the chimpanzee as having erect posture, Blumenbach proposed "that apes should not be seen as either quadrupeds or bipeds but as *Quadrumana* (four-handed)" (85).

25. Cornell likens this upward-striving tendency to "Faustian excess" ("Faustian," 485).

26. Although some readers of Lamarck, including Darwin, believed that his works contained a support of volition, Mayr argues that such readings are largely a result of a mistranslation of terms. Lamarck, according to Mayr, was completely opposed to any vitalistic or teleological explanations of change (357).

27. For a detailed account of this debate from Darwin's time to our own, see Wenzel ("Goethe und Darwin," 145–58). Although Haeckel is the better known proponent, Wenzel traces the argument back to Rudolf Virchow's publication in 1861, "Göthe als Naturforscher und in besonderer Beziehung auf Schiller" (148).

28. What is surprising about their respective arguments is that each to a large degree admits that Goethe's thought contains seeds of evolutionary thought and language. Although D. Kuhn writes that Herder (in collaboration with Goethe) suggested something akin to the descent of the species in his *Ideas Concerning the Philosophy of the History of Mankind* (for which Kant criticized the work), she argues that Goethe did not embrace a theory of descent (10–12) and that he abandoned even

thinking about it after his return from Italy (14). Her argument that "the barrier of Christian dogma must have been too prohibitive" (12) might explain why Goethe did not publish certain works (13) or why he was very careful in his presentation of the topic, but it does not seem an adequate explanation as to why Goethe would not have researched the idea on his own—especially since his writings—published and nonpublished—questioned other popular, Christian notions such as teleology, divine intervention in the natural order, and creation ex nihilo. Indeed, the fact that Goethe omitted commenting upon evolutionary matters in the works of others (14–15) may suggest that he viewed evolution as a sensitive political issue. While I agree with Kuhn that the appearance of phenomena took precedence over other issues for Goethe (14), I do not believe that this primacy of interest necessarily rules out an interest in evolution. Wells, too, suggests that it was impossible for Goethe or his contemporaries "to grasp the (to us) so straightforward and obvious principle that widely different plants and animals have common ancestors" (1). He, however, also presents numerous citations from Goethe's works "that might appear to commit him wholeheartedly to evolution" (28). He even argues that Goethe made statements to support the belief in common ancestry for related species, but that these statements "do not imply that the principle of descent may be given a general application to the whole animal kingdom" (30). Wenzel, like Wells, argues that Goethe limited his theories of morphological change to the vertebrates. He maintains, however, any true evolutionary theory would provide a model for all life-forms, not merely for the vertebrates (56). Wenzel, like Kuhn and Wells, concludes that Goethe was a typical representative of his time: he had begun to abandon Linnaean rigidity and argue for more fluidity of form, but had not yet embraced Darwinian evolutionary thinking (62). In contrast to these scholars, Bräuning-Oktavio argues that Goethe's essay on the intermaxillary bone, taken together with his personal correspondence, demonstrates that Goethe did indeed believe in the descent of the species (43–53). In particular, Bräuning-Oktavio cites Goethe's letter to Knebel (Weimar, 17 November 1784) to support his point. Goethe writes, "Ich habe mich enthalten das Resultat, worauf schon Herder in seinen Ideen deutet, schon ietzo mercken zu lassen, daß man nämlich den Unterschied des Menschen vom Thier in nichts einzelnem finden könne. Vielmehr ist der Mensch aufs nächste mit den Thieren verwandt . . ." (FA, 2,2: 553). Charlotte von Stein, in a letter to Knebel, goes so far as to state that Herder's work makes apparent "daß wir erst Pflanzen und Tiere waren" and that Goethe was currently occupying himself with these notions (1 May 1784, cited in Bräuning-Oktavio, 53).

29. Foucault does not mention Goethe directly, but speaks generally of scientists of that period.

30. Lenoir examines the connection between Goethe and Darwin and argues that the main difference between the two men was that "Goethe's biology was fundamentally and radically teleological in character" ("Eternal," 18). He contends that Goethean organisms change and adapt due to the conditions of environment, while Darwinian ones "are far more passive and less tenacious in their grip on life:

they simply vary—spontaneously" (27). Lenoir, however, in emphasizing the environment's role in influencing changes, does not discuss the animal's own will as part of that change. Like Lenoir, Cornell argues that Goethean organisms were much more active than Darwinian ones. Cornell further argues that Goethe's organisms do not change due to "hidden purposes of hereditary elements—such as genes." Instead, Cornell stresses that "the animals' visible strivings are essential elements of the evolutionary logic" ("Faustian," 487). Cornell argues that for Goethe "the question of modern biology lies beyond the subject of evolution per se. For him the question is not so much whether, or by what great ancestors, we are kin with other organisms, but how to think about man's odd place in nature, and about man as a knower of nature, while recognizing that kinship" ("Faustian," 489).

31. Goethe does not state in his own voice the most radical aspects of evolution—i.e., that species have progressed over time and have become more specialized—but he quotes another scientist. His agreement with the theory, however, is evident. He views an ancient steer, whose fossilized skeleton he studied, as a "Stamm-Race" from which the common steer is a descendant (Abkömmling [FA, I, 24: 553]).

32. This progression of life evolving from the sea follows the suggestions made in *Faust II*. The manufactured spirit, Homunculus, searches for a way to be born into flesh. He is advised by Proteus to be born from the sea as that is where simple life evolves into more complex forms:

> you must begin out in the open sea
> That is where you start on a small scale,
> glad to ingest the smallest creatures;
> little by little you'll increase in size
> and put yourself in shape for loftier achievements.
> (2: 210)

> Im weiten Meere mußt du anbeginnen!
> Da fängt man erst im Kleinen an
> Und freut sich Kleinste zu verschlingen,
> Man wächst so nach und nach heran,
> Und bildet sich zu höherem Vollbringen.
> (8260–64)

33. Goethe mentions several times how other scientists wrote in code or anagrammatically to hide some of their meaning from less educated or attentive readers. See, for example, "Erfinden and Entdecken," where Goethe discusses how Galileo was able to "hide his discovery anagrammatically in Latin verse" [versteckte seine Erfindung anagrammatisch in lateinische Verse] and where he speaks of this method as of an "open secret" (dieses öffentlichen Geheimnisses [FA, I, 25: 38]). See also his discussion of Descartes's fears in light of Galileo's troubles (FA, I, 23, no. 1: 709).

34. See LA (2, 10A: 899) and Wells (30). In "Morphologie," Goethe states, "Die Gestalt ist ein bewegliches, ein werdendes, ein vergehendes. Gestaltenlehre ist Verwandlungslehre. Die Lehre der Metamorphose ist der Schlüssel zu allen Zeichen der Natur" (FA, 1, 24: 349).

35. Some believe that Goethe is as extreme in his praise of Howard as he is in his criticisms of Newton (Martin, 183, 185ff.).

36. In his poem, "In Honour of Howard," Goethe even repeats the language of actions and passions in reference to the nimbus cloud: "Der Erde tätig-leidendes Geschick" (FA, 1, 25: 240). Martin argues that Goethe is here inspired by Platonic philosophy (186).

37. Martin, however, argues that, because of the use of "tätig-leidendes" in the Nimbus verse of "In Honour of Howard," this cloud is of special importance to Goethe. He further believes that this passage bears significance due to a reference to Plato's 7th Letter (193–94).

38. According to Goethe, these are the clouds in India that cross the land in unending changes of form (unendlicher Gestaltveränderung). He further explains that Kalidasa speaks of these clouds in his epic poem *The Cloud Messenger.*

39. Goethe cites this poem as an inspiration to both his meteorological studies and his poem "In Honour of Howard."

40. Goethe here quotes the English from Wilson's translation of Kalidasa's poem.

41. These lines are from the 1821 English translation (by Bowring) for *Gold's London Magazine* (cited here from FA, 1, 25: 239).

42. In the end, Faust does not actually desire time to stand still, but can imagine in theory the conditions necessary for him to do so (11,574–86).

43. In his "Fossil Steer," Goethe similarly explains that animal horns appear beautiful because their curves imply movement although they are actually at rest. He cites Hogarth's aesthetic theory as being in agreement with his own observations of nature (FA, 1, 24: 558).

44. "Erster Entwurf" and "Vorträge, über die drei ersten Kapitel des Entwurfs" were written in 1795 and 1796, respectively.

45. Here, one wonders whether Goethe's repeated use of the concept of resignation in his literary works, from *Wilhelm Meister's Journeyman Years* to *Elective Affinities,* is not related to the principle of compensation as well. In a way, those characters who practice renunciation, including Wilhelm or Charlotte, attempt to limit themselves to attain higher goals.

Chapter 4

1. The names within the poem, as well as the themes of love, are taken from Theocritus's *Idylls* (11th) and Vergil's *Eclogues.*

2. Traditionally, the marriage between the elm and vine was considered ideal because, once the vine has the supportive structure of the elm, it will be more productive and capable of bringing forth an abundance of grapes. Froebe argues that

Goethe quite possibly saw examples of this method of growing grapes (175). Demetz traces this positive image of their relationship through such authors as Catullus, Horace, Vergil, Cato, and Milton.

3. Goethe explores a similar theme in the poem "The Bride from Corinth" ("Die Braut von Korinth"), where the feminine character sucks the lifeblood from her lover.

4. Roe further argues that it is "significant, with regard to their utilization of attractive forces, that both Maupertuis and Buffon were early supporters of Newton on the Continent. The freedom that Newtonian mechanism offered to biology—the addition of force to matter and motion as the fundamental categories of explanation—played an important role in embryology as well as in other areas of physiology" (18).

5. And, as already discussed in chapter 3, many of the most prominent pre-Darwinian theories, as well as Darwin's own, have strong Newtonian influences. Even in the Cuvier-Geoffroy debate, Cuvier was generally considered to have been triumphant over Geoffroy.

6. As already discussed in chapter 2, Goethe criticized the epigenetic theories of Wolff as being too based on material principles

7. Nisbet traces the idea of "an integral whole" consisting "itself of units which are also integral wholes" to Bonnet and to the earlier Platonic and Leibnizian traditions (18–20). Nisbet argues that Goethe's theories were different from "earlier equivalents" in the "mass of empirical evidence, the carefully accumulated observations he adduces in their support" (18).

8. Goethe's approach in his botany of attempting to look at the whole organism as opposed to breaking it up into parts departs radically from the approach of Linnaeus (see Waenerberg, 30).

9. In a somewhat similar spirit, Lewontin writes that biologists working today according to the models of Descartes have tended to oversimplify organic processes. By predominantly focusing on DNA (as the predetermined and fixed aspect), these biologists miss other important factors that influence the end result of an organism (70–73). He explains, "If we had the complete DNA sequence of an organism and unlimited computational power, we could not compute the organism, because the organism does not compute itself from its genes . . . There exists, and has existed for a long time a large body of evidence that demonstrates that the ontogeny of an organism is the consequence of a unique interaction between the genes that it carries, the temporal sequence of external environments through which it passes during its life, and random events of molecular interactions within individual cells. It is these interactions that must be incorporated into any account of how an organism is formed" (17–18). One of the examples that he gives to illustrate his point is that of the tropical vine, *Syngonium*. The structure of this plant, the shape of its leaves, and even whether it is positively or negatively geotropic, varies tremendously in different environments—environments that it, in a way, seeks out through its own vines.

10. Goethe similarly discusses the two modes of reproduction in "The Purpose Set Forth": "The above axiom concerning the coexistence of multiple identical and similar entities leads to two further cardinal principles of the organism: propagation by bud and propagation by seed. In fact these principles are simply two ways of expressing the same axiom. We will seek to trace these two paths through the entire realm of organic nature and in the process will find that many things fall vividly into place" (12: 65) [Gemmation und Prolifikation sind abermals zwei Hauptmaximen des Organismus, die aus jenem Hauptsatz der Koexistenz mehrer gleichen und ähnlichen Wesen sich herschreiben und eigentlich jene nur auf doppelte Weise aussprechen. Wir werden diese beiden Wege durch das ganze organische Reich durchzuführen suchen, wodurch sich manches auf eine höchst anschauliche Weise reihen und ordnen wird] (FA, 1, 24: 395).

11. Similarly, her affinities with a sterile flower, the aster, emphasize this point.

12. Significantly, many believed that the sperm played no role in reproduction (Farley, 20–21). Linnaeus, in contrast, argued that both sexual organs had to be present for reproduction to occur (Farley, 23).

13. Goethe explicitly makes the connection between politics and botany when discussing Rousseau's botanical works (FA, 1, 24: 741–42).

14. Reproduction, sexuality, and gender roles were central topics of discussion in the eighteenth and early nineteenth centuries. Debates raged between the epigenesists and the preformationists in this period. Scientists and amateur botanists became increasingly interested in plant sexuality. In the late seventeenth century, scientists had discovered that plants reproduce sexually, and the eighteenth century struggled to understand the roles of the sexes in reproduction. (For an extensive discussion of the history of plant sexuality, see Schiebinger, 18–39.) Technological advances assisted this study. Eighteenth-century scientists, like Buffon, used microscopes to search for the secrets of reproduction (Sloan, 415–34). Linnaeus developed his classification of plants around sexual difference (male parts determined the class of the plants, female parts its order), and he (as Erasmus Darwin in England) used the anthropomorphic language of marriage to describe plant reproduction. The interest in plants was not isolated to the academy, but also extended to the middle class. Botany was one of the few sciences that women were encouraged to practice (Schiebinger, 36; Shteir), and both Rousseau and Goethe encouraged women to study plants and wrote didactic pieces to educate them to do so.

15. These barnacles were called goose barnacles because up until the eighteenth century it was believed that they were actually the eggs of barnacle geese (Anderson, 1–2). Linnaeus first classified barnacles and included the multivalve shells in a single genus. Linnaeus viewed barnacles as mollusks, and this belief was followed by Cuvier, Buffon, and Lamarck. In 1829, the British army surgeon J. Vaughan Thompson proved their relationship to crustaceans. Lamarck subdivided the barnacles into two orders (Sessile and Pedunculated) (see Anderson, 2).

16. Gray, for instance, offers a complicated explanation for Goethe's emphasis on the similarity, rather on the polar nature, of the sexes. He turns to Goethe's

notebooks to show that Goethe initially "tended to think that the male and the fe- male reproductive organs each represented one of these opposed tendencies" (86). He claims, however, that since Goethe was wedded to the alchemical ideal of seven phases, he was forced to collapse two stages of plant development into one (87).

17. "In many instances the style looks almost like a filament *without anthers;* the two resemble one another in external form more than any of the other parts" (12: 87, emphasis added) [In vielen Fällen sieht der Griffel fast einem Staubfaden *ohne Anthere gleich,* und die Verwandtschaft ihrer Bildung is äußerlich größer als bei den übrigen Teilen] (no. 69, emphasis added).

18. Two different essays were published under the title "Über die Spiral- tendenz der Vegetation." The first was published in 1831 in "Versuch über die Metamorphose der Pflanzen" (Stuttgart: Cotta'sche Buchhandlung). The second and longer, but more aphoristic, essay was published posthumously in 1833 in *Goethes nachgelassene Werke* ([Stuttgart: Cotta'sche Buchhandlung], vol. 50). Unless otherwise noted, the references are to the essay of 1833.

19. The reaction to Goethe's social and scientific novel, *Elective Affinities,* is telling of his ambiguous moral relationship to his own time. Then, as now, there is no critical agreement whether the novel is conservative or radical, i.e., whether it praises or condemns the institution of marriage or whether it supports or criti- cizes the sexual practices of the Jena Romantics.

20. Similarly, "[o]ne system cannot be imagined apart from the other, for only the two working together can achieve a vital effect" (Mueller translation, 137) [keins kann von dem andern abgesondert gedacht werden, weil eins durch das an- dere nur lebendig wirkt] (FA, 1, 24: 794–95).

21. For discussions of this symbolism, see Burckhardt (130ff.). Froebe discusses at length the reconciliatory symbolism behind this hermetic image. Mercury, the god who carried this staff, is often seen as a mediator between the gods and hu- man beings as well as a mediator between the sun and the moon (181–88)

22. Jung points to the obvious reference of this alchemical symbol of the snake and tree to the biblical story of the Garden of Eden. He explains the significance of the snake/tree symbolism as representing the life force of the tree. The snake is the "chthonic spiritus vegetativus" and the "arcane substance transforms itself in- side the tree and thus constitutes its life" (315).

23. Froebe also argues that this image is symbolic for enduring friendships (176).

24. Demetz also reports that the elm and the vine were used in early Christian symbolism to represent the mystical wedding of the rich and the poor: "the first, voluntarily offering the riches of God has granted them to sustain the spiritually stronger yet materially weaker; the latter, voluntarily sacrificing the energy of their prayers to sustain the materially more privileged yet spiritually less gifted group of the Christian community" (526). It is interesting to note that in Milton's descrip- tion of Eve, her hair is likened to vines and takes on a distinct sexual quality:

> Shee as a vail down to the slender waste
> Her unadorned gold'n tresses wore
> Dissheveld, but in wanton ringlets wav'd
> As the Vine curles her tendrils, which impli'd
> Subjection, but requir'd with gentle sway,
> And by her yielded, by him best receivd
> Yielded with coy submission, modest pride,
> And sweet reluctant amorous delay.
> (*Paradise Lost*, IV, 304–11)

25. Flax has long been a symbol of masculinity and represented by Old European gods, such as the Lithuanian Vaižgantas, who is "worshipped by women, born from the earth, tortured as flax is tortured, who died and was resurrected next spring" (Gimbutas, 19–25).

26. The term is from Haller's treatise, *De partibus corporis humani sensilibus et irritabilibus* (1753), which was translated into German as *Von den empfindlichen und reizbaren Teilen des menschlichen Körpers*. For a discussion on how the medical term *Reizbarkeit* became important in German aesthetic discussions, see Richter (esp. 97–102).

27. Feminist care ethicists use similar language to describe the differences between the sexes. For example, Gilligan quotes Chodorow to describe how the masculine personality develops to stand by itself, whereas the feminine one "comes to define itself in relation and connection to other people more than masculine personality does" (7). While the emphasis here is on interrelatedness to others generally and not to the masculine specifically, the point is that the feminine personality is not as independent as the masculine one. Both Noddings and Gilligan focus on the masculine moral rigidity, with its adherence to set structures and rules, while pointing to the more fluid feminine propensities of relatedness, nurturing, and responsiveness as evidenced by the feminine care ethic (Gilligan, 2, 48, 73, 164–65, 175; Noddings, *Women*, 62–63).

28. Froebe compares the eighteenth-century interest in spiral vessels to the interest in genetic codes today (166).

29. It is perhaps important to emphasize at this point that Goethe is not speaking of female plants as having greater freedom of movement, but plants (whether male or female) that possess the feminine tendency. In other words, it is not a question of an individual of a particular sex being more free, but an individual who, regardless of sex, possesses certain characteristics. Therefore, a plant that is sexually male could possess, in Goethe's schema, predominantly feminine characteristics. Similarly, if one were to turn to Goethe's literary works in order to analyze his characters according to method that he outlines in his botanical writings, one would examine his male and female characters according to both "masculine" and "feminine" tendencies.

30. "Whether these first beginnings could be conclusively traced in opposing directions, to the plant through light and to the animal through darkness, I do not

make bold to decide, although opinions and analogies are not lacking on this subject" (Mueller translation, 25) [Ob diese ersten Anfänge, nach beiden Seiten determinabel, durch Licht zur Pflanze, durch Finsternis zum Tier hinüber zu führen sind, getrauen wir uns nicht zu entscheiden, ob es gleich hierüber an Bermerkungen und Analogie nicht fehlt] (FA, 1, 24: 394).

31. Goethe wrote "Amyntas" in September of 1797 and the "Metamorphosis" poem in June of 1798. Both poems were published in the same journal in 1799 and again in Goethe's *Neue Schriften* in 1800. In *Neue Schriften,* "Amyntas" is followed by "Metamorphosis," which is then followed in turn by "Hermann und Dorothea." He then also published "Metamorphosis" in 1817 within an autobiographical, scientific essay, "Fate of the Printed Work" ("Schicksal der Druckschrift").

32. Goethe also praises Rousseau's attempts to teach botany to women (FA, 1, 24: 743)

33. Koerner argues that "[a]s the poem closes, it becomes clear that he [Goethe] aims to make the female reader simultaneously a passive recipient and an object of knowledge." The last lines of the poem, however, stress equality: a harmony of mind and perception. Koerner similarly wishes to argue that Goethe's poem "Gefunden" ("Found") is part of "his seducer's repertoire" and exhibits Goethe's "concept that the male voice is dependent on women's silence" (494). The poem, given to Christiane Vulpius, describes how the poet (Goethe) discovers a flower (Vulpius) in the woods and transplants it in his garden. According to Koerner's interpretation, Goethe "trades material protection" for "ongoing sexual favors" and expects silence in return: "He makes her what we most fear becoming: a vegetable" (494). Koerner's interpretation of the poem, however, appears to rest upon a mistranslation of it. *Zweigen* (from the German noun *Zweig*—branch, hence meaning to branch out, to sprout, to grow) becomes *silence* (I presume from schweigen—silence). "Nun zweigt es immer / Und blüht so fort," which should describe a kind of flourishing and blossoming, becomes "Now she is always silent / And continues to bloom" (494).

34. In his poem "Blessed Longing" ("Selige Sehnsucht"), Goethe similarly speaks of a higher mating or union (höherer Begattung).

35. These authors do not reject the division of masculine and feminine characteristics, but demand that we recognize the contributions of the feminine as different, yet equal.

36. Butler admonishes that "it is no longer possible to take anatomy as a stable referent" when discussing gender (*Bodies,* 63). I would argue, however, that Goethe's view of natural gender differs to some extent from the paradigm that Butler criticizes: first, he views nature itself as fluid, not stable; and, second, he is conscious that we may impose too much of ourselves upon our observations. For Butler's criticisms of a stable notion of gender, see, for example, *Gender* (esp. 1–10, 30). Of course, in the end, nature for Butler remains primarily a culturally created construct, where for Goethe nature is both outside of culture while also a part of it. So he tells us in his *Theory of Colors* that when we study nature, we must do so

self-consciously and with irony, because we too are both part of nature and out-side of it. He warns that we theorize (theoretisieren) "by every observing look at the world" [bei jedem aufmerksamen Blick in die Welt] (FA, 1, 23, pt. 1: 14). Goethe similarly asks in a maxim, "Is it the object or is it you who is here express-ing itself?" [Ist es der Gegenstand oder bist du es, der sich hier ausspricht] ("Maximen und Reflexionen," FA, 1, 13: 50, no. 1.320). Butler also believes that we need to be self-conscious in our examination, but where for Goethe such irony and self-consciousness may lead to an evaluation of nature, for Butler this leads to a questioning of the validity of basing anything upon nature at all: "Only from a self-consciously denaturalized position can we see how the appearance of natural-ness is itself constituted" (Gender, 110). Although both Goethe and Butler distance the discussion of gender from sexual organs, for Goethe this means that one can still talk about masculine and feminine qualities or tendencies, whereas for Butler this means we constantly need to subvert such categories.

Conclusion

1. It is interesting to note that Lewontin, in his recent book on modern biology, raises a similar point. Although Lewontin is not a vitalist and his scientific pro-gram is different from Goethe's, he too stresses that current science needs to reevaluate its Cartesian underpinnings. He argues that Descartes's clock model "has led to an overly simplified view of the relations of parts to wholes and of causes to effects" (72). Descartes's model, moreover, "is not only a description of how the world operates but also a manifesto for how to study natural phenomena" (73). This manifesto, in his eyes, has narrowed the sphere of scientific inquiry: "Science as we practice it solves those problems for which its methods and con-cepts are adequate, and successful scientists soon learn to pose only those prob-lems that are likely to be solved" (73). Throughout his book, he argues that more attention needs to be paid to organisms as active instead of passive entities, espe-cially in their ability to create their own environments. He also argues that scien-tists need to look to a complex set of factors outside of DNA to understand how organisms develop.

2. Goethe was notorious for his inability to take criticism—either about his lit-erature or his scientific works. He was especially sensitive to criticism on his Theory of Colors (as Eckermann and Schopenhauer discovered) and resented any chal-lenges to it.

3. To take a later example from the nineteenth century, Darwin could put aside questions about the causes of variation, because the principle of natural selection works whether we know the causes of the variations or not (Behe, 6; Mayr, 68–83, 833).

BIBLIOGRAPHY

Editions of Goethe's Works Cited

Goethe, Johann Wolfgang von. *Sämtliche Werke.* Ed. Hendrik Birus et al. 40 vols. Frankfurt: Deutscher Klassiker, 1985–. (Abbreviated in the text as FA.)

————. *Goethes Werke.* Ed. Erich Trunz. 9th ed. 14 vols. Hamburg: Christian Wegner, 1950–68. (Abbreviated in the text as HA.)

————. *Gedenkausgabe der Werke, Briefe und Gespräche.* Ed. Ernst Beutler. Artemis Ausgabe. 24 vols. Zurich: Artemis, 1948–54. (Abbreviated in the text as GA.)

————. *Goethes Gespräche.* Ed. Wolfgang Herwig. 5 vols. Zurich: Artemis, 1965–87.

————. *Die Schriften zur Naturwissenschaft.* Ed. Rupprecht Mattheai, Wilhelm Troll, and K. Lothar Wolf. 2 parts. 11 vols. Weimar: Hermann Böhlaus Nachfolger, 1947–. (Abbreviated in the text as LA.)

————. *Goethe's Collected Works.* 12 vols. New York: Suhrkamp, 1983–89.

————. *Conversations with Eckermann.* Trans. John Oxenford. San Francisco: North Point, 1984.

————. *Goethe's Botanical Writings.* Trans. Bertha Mueller. Woodbridge, Conn.: Ox Bow, 1989.

Works Cited

Adler, Hans. "Erfahrung." *Goethe Handbuch.* Vol. 4, no. 1. Ed. Bernd Witte et al. Stuttgart: J. B. Metzler, 1998: 272–74.

————. "Erkenntnis." *Goethe Handbuch.* Vol. 4, no. 1. Ed. Bernd Witte et al. Stuttgart: J. B. Metzler, 1998: 277–80.

Adler, Jeremy. *"Eine fast magische Anziehungskraft": Goethes Wahlverwandtschaften und die Chemie seiner Zeit.* Munich: C. H. Beck, 1987.

Altner, Günter. "Goethe as a Forerunner of Alternative Science." *Goethe and the Sciences: A Reappraisal.* Ed. Frederick Amrine, Francis J. Zucker, and Harvey Wheeler. Boston Studies in the Philosophy of Science, vol. 97. Boston: D. Reidel, 1987: 341–50.

d'Alton, E[duard Joseph]. *Das Riesen-Faulthier, Bradypus giganteus, abgebildet, beschrieben, und mit den verwandten Geschlechtern verglichen.* Bonn: In Commission bei Eduard Weber, 1821.

————. *Die Skelete der Nagethiere, abgebildet und verglichen.* Bonn: In Commission bei Eduard Weber, 1823.

Anderson, D. T. *Barnacles: Structure, Function, Development and Evolution.* London: Chapman and Hall, 1994.

Appel, Toby A. *The Cuvier-Geoffroy Debate: French Biology in the Decades Before Darwin.* Oxford: Oxford University Press, 1987.

Arber, Agnes. "Goethe's Botany." *Chronica botanica* 10 (1946): 63–124.

Baier, Annette C. *Moral Prejudices: Essays on Ethics.* Cambridge: Harvard University Press, 1994.

Barsanti, Giulio. "Lamarck and the Birth of Biology 1740–1810." *Romanticism in Science: Science in Europe, 1790–1840.* Ed. Stefano Poggi and Maurizio Bossi. Dordrecht: Kluwer Academic Publishers, 1994: 47–74.

Behe, Michael J. *Darwin's Black Box: The Biochemical Challenge to Evolution.* New York: Simon and Schuster, 1996.

Bell, Matthew. *Goethe's Naturalistic Anthropology: Man and Other Plants.* Oxford: Clarendon, 1994.

Benjamin, Walter. *Selected Writings.* Vol. 1: 1913–1926. Ed. Marcus Bullock and Michael W. Jennings. Cambridge: Harvard University Press, 1996.

Bennett, Benjamin. *Goethe's Theory of Poetry: Faust and the Regeneration of Language.* Ithaca: Cornell University Press, 1986.

Bielschowsky, Albert. *The Life of Goethe.* Trans. William A. Cooper. Vol. 2. New York: G. P. Putnam's Sons, 1907.

Bortoft, Henri. *The Wholeness of Nature: Goethe's Way Toward a Science of Conscious Participation in Nature.* Hudson, N.Y.: Lindisfarne, 1996.

Boyle, Nicholas. *Goethe: The Poet and the Age.* Vol. 1. Oxford: Clarendon, 1991.

Brady, Ronald. "Form and Cause in Goethe's Morphology." *Goethe and the Sciences: A Reappraisal.* Ed. Frederick Amrine, Francis J. Zucker, and Harvey Wheeler. Boston Studies in the Philosophy of Science, vol. 97. Boston: D. Reidel, 1987: 257–300.

Bräunig-Oktavio, Hermann. *Vom Zwischenkieferknochen zur Idee des Typus: Goethe als Naturforscher in den Jahren 1780–1786.* Nova Acta Leopoldina. New Series, no. 126, vol. 18. Leipzig: Johann Ambrosius Barth, 1956.

Bricmont, J. "Science of Chaos or Chaos in Science?" *The Flight From Science and Reason.* Ed. Paul R. Gross, Norman Levitt, and Martin W. Lewis. New York: New York Academy of Sciences, 1996: 131–75.

Brodsky, Claudia. "The Colouring of Relations: *Die Wahlverwandtschaften* as Farbenlehre." *Modern Language Notes* 97 (1982): 1147–79.

Burckhardt, Titus. *Alchemy: Science of the Cosmos, Science of the Soul.* Trans. William Stoddart. London: Stuart and Watkins, 1967.

Burgard, Peter. *Idioms of Uncertainty: Goethe and the Essay.* University Park: Pennsylvania State University Press, 1992.

Burwick, Frederick. *The Damnation of Newton: Goethe's Color Theory and Romantic Perception.* New York: Walter de Gruyter, 1986.

Butler, Judith. *Bodies That Matter: On the Discursive Limits of "Sex".* New York: Routledge, 1993.

———. *Gender Trouble: Feminism and the Subversion of Identity.* New York: Routledge, 1990.

Clarke, Desmond. "Descartes' Philosophy of Science and the Scientific Revolution." *The Cambridge Companion to Descartes.* Ed. John Cottingham. Cambridge: Cambridge University Press, 1992: 258–85.

Coen, Enrico S., and Rosemary Carpenter. "The Metamorphosis of Flowers." *Plant Cell* 5 (1993): 1175–81.

Cornell, John F. "Analogy and Technology in Darwin's Vision of Nature." *Journal of the History of Biology* 17 (1984): 303–44.

———. "Faustian Phenomena: Teleology in Goethe's Interpretation of Plants and Animals." *Journal of Medicine and Philosophy* 15 (1990): 481–92.

———. "Goethe on Plants, Animals and Modern Biologists." *St. John's Review* 63 (1995): 39–57.

———. "Newton of the Grassblade? Darwin and the Problem of Organic Teleology." *Isis* 77 (1986): 405–21.

Demetz, Peter. "The Elm and the Vine: Notes Toward the History of the Marriage Topos." *PMLA* 73 (1958): 521–32.

Eissler, Kurt. *Goethe: A Psychoanalytic Study 1775–1786*. Vol. 2. Detroit: Wayne State University Press, 1963.

Engelhardt, Dietrich von. "Historical Consciousness in the German Romantic *Naturforschung*." Trans. Christine Salazar. *Romanticism and the Sciences*. Ed. Andrew Cunningham and Nicholas Jardine. Cambridge: Cambridge University Press, 1990: 56–67.

Erpenbeck, John. "'. . . die Gegenstände der Natur an sich selbst . . .': Subjekt und Objekt in Goethes naturwissenschaftlichem Denken seit der italienischen Reise." *Goethe Jahrbuch* 105 (1988): 212–33.

———. "Wissenschaft." *Goethe Handbuch*. Vol. 4, no. 2. Ed. Bernd Witte et al. Stuttgart: J. B. Metzler, 1998: 1187–94.

Farley, John. *Gametes and Spores: Ideas about Sexual Reproduction 1750–1914*. Baltimore: Johns Hopkins University Press, 1982.

Fink, Karl J. *Goethe's History of Science*. Cambridge: Cambridge University Press, 1991.

Finley, Gerald. "Pigment into Light: Turner, and Goethe's *Theory of Colors*." *European Romantic Review* 2 (1991): 39–60.

Flannery, Maura C. "Goethe and Arber: Unity in Diversity." *American Biology Teacher* 57 (1995): 544–47.

Fleck, Ludwik. *Genesis and Development of a Scientific Fact*. Chicago: University of Chicago Press, 1979.

Foucault, Michel. *The Order of Things*. New York: Vintage, 1973.

Friedrichsmeyer, Sara. *The Androgyne in Early German Romanticism: Friedrich Schlegel, Novalis and the Metaphysics of Love*. Stanford German Studies. New York: Peter Lang, 1983.

Froebe, Hans A. "'Ulmbaum und Rebe': Naturwissenschaft, Alchymie und Emblematik in Goethes Aufsatz 'Über die Spiraltendenz' (1830–1831)." *Jahrbuch des Freien Deutschen Hochstifts* 1969: 164–93.

Gilli, Marita. "Das Verschweigen der Geschichte in Goethes *Wahlverwandtschaften* oder Wie man der Geschichte nicht entfliehen kann." *Sie, und nicht Wir*. Ed.

Arno Herzig, Inge Stephan, and Hans G. Winter. Hamburg: Doelling und Galitz, 1989: 553–65.

Gilligan, Carol. *In a Different Voice: Psychological Theory and Women's Development.* Cambridge: Harvard University Press, 1982.

Gimbutas, Marija. "Pre–Indo-European Goddesses in Baltic Mythology." *Mankind Quarterly* 26 (1985): 19–25.

Gleick, James. *Chaos: Making a New Science.* New York: Penguin, 1988.

Gode-von Aesch, Alexander Gottfried Friedrich. *Natural Science in German Romanticism.* New York: Columbia University Press, 1941.

Gögelein, Christoph. *Zu Goethes Begriff von Wissenschaft: Auf dem Wege der Methodik seiner Farbstudien.* Munich: Carl Hanser, 1972.

Göres, Jörn. "Polarität und Harmonie bei Goethe." *Deutsche Literatur zur Zeit der Klassik.* Ed. Karl Otto Conrady. Stuttgart: Reclam, 1977: 93–113.

Goldstein, Sheldon. "Quantum Philosophy: The Flight from Reason in Science." *The Flight From Science and Reason.* Ed. Paul R. Gross, Norman Levitt, and Martin W. Lewis. New York: New York Academy of Sciences, 1996: 119–25.

Graham, Ilse. *Goethe: Portrait of the Artist.* New York: Walter de Gruyter, 1977.

Gray, Ronald. *Goethe the Alchemist.* Cambridge: Cambridge University Press, 1952.

Gross, Paul R., Norman Levitt, and Martin W. Lewis, eds. *The Flight From Science and Reason.* New York: New York Academy of Sciences, 1996.

Gundolf, Friedrich. *Goethe.* Berlin: Bondi, 1920.

Haeckel, Ernst. *Natürliche Schöpfungsgeschichte.* Berlin: Georg Reimer, 1868.

Heisenberg, Werner. "Die Goethe'sche und die Newton'sche Farbenlehre im Lichte der modernen Physik." *Geist der Zeit* 19 (1941): 261–75. Quoted from Heisenberg's *Gesammelte Werke/Collected Works.* Series C: Philosophical and Popular Writings, vol. 1. Munich: Piper, 1984: 146–60.

———. "Das Naturbild Goethes und die technisch-naturwissenschaftliche Welt." *Goethe: Neue Folge des Jahrbuchs der Goethe-Gesellschaft* 29 (1967): 27–42. Quoted from Heisenberg's *Gesammelte Werke/Collected Works.* Series C: Philosophical and Popular Writings, vol. 2. Munich: Piper, 1984: 394–409.

———. "The Teachings of Goethe and Newton on Colour in the Light of Modern Physics." Translation of "Die Goethe'sche und die Newton'sche Farbenlehre im Lichte der modernen Physik." *Philosophic Problems of Nuclear Science.* Trans. F. C. Hayes. New York: Pantheon, 1952: 60–76.

———. "Tradition in Science." *Science and Public Affairs: Bulletin of the Atomic Scientists* 29 (1973): 4–10. Quoted from Heisenberg's *Gesammelte Werke/Collected Works.* Series C: Philosophical and Popular Writings, vol. 3. Munich: Piper, 1985: 440–46.

Heitler, Walter. "Die Naturwissenschaft Goethes." *Naturphilosophische Streifzüge.* Braunschweig, Germany: Friedrich Vieweg und Sohn, 1970: 66–76.

Helmholtz, Hermann von. "Goethe's Presentiments of Coming Scientific Ideas." *Science and Culture: Popular and Philosophical Essays.* Ed. David Cahan. Chicago: University of Chicago, 1995: 393–412.

————. "On Goethe's Scientific Researches." *Science and Culture: Popular and Philosophical Essays*. Ed. David Cahan. Chicago: University of Chicago, 1995: 1–17.

Höpfner, Felix. "'Wirkungen werden wir gewahr [. . .]': Goethes *Farbenlehre* im Widerstreit der Meinungen." *Goethe Jahrbuch* 111 (1994): 203–11.

————. *Wissenschaft wider die Zeit: Goethes Farbenlehre aus rezeptionsgeschichtlicher Sicht*. Heidelberg: Carl Winter Universitätsverlag, 1990.

Howells, Bernard. "The Problem with Colour. Three Theorists: Goethe, Schopenhauer, and Chevreul." *Artistic Relations: Literature and the Visual Arts in Nineteenth-Century France*. Ed. Peter Collier and Robert Lethbridge. New Haven: Yale University Press, 1994: 76–93.

Huber, Peter. "Polarität/Steigerung." *Goethe Handbuch*. Vol. 4, no. 2. Ed. Bernd Witte et al. Stuttgart: J. B. Metzler, 1998: 863–65.

Jackson, Myles. "Natural and Artificial Budgets: Accounting for Goethe's Economy of Nature." *Science in Context* 3 (1994): 409–31.

Jahn, Ilse. "On the Origin of Romantic Biology and Its Further Development at the University of Jena Between 1790 and 1850." *Romanticism in Science: Science in Europe 1790–1840*. Ed. Stefano Poggi and Maurizio Bossi. Dordrecht: Kluwer Academic Publishers, 1994: 75–89.

Jardine, Nicholas. "*Naturphilosophie* and the Kingdoms of Nature." *Cultures of Natural History*. Ed. N. Jardine, J. A. Secord, and E. C. Spray. Cambridge: Cambridge University Press, 1996: 230–45.

Jung, Carl. *Alchemical Studies*. Vol. 13 of *The Collected Works*. Ed. Herbert Read et al. Trans. R. F. C. Hull. Princeton: Princeton University Press, 1977.

Kirby, Margaret. "Classical Science?" *A Reassessment of Weimar Classicism*. Ed. Gerhart Hoffmeister. Lewiston: Edwin Mellen, 1996: 64–76.

Klumbies, Gerhard. "Die Weiterentwicklung der vergleichenden Betrachtungsweise Goethes in der Anatomie, Physiologie und Verhaltensforschung." *Goethe und die Wissenschaften*. Ed. Helmut Brandt. Jena: Friedrich-Schiller-Universität, 1984: 52–58.

Koerner, Lisbet. "Goethe's Botany: Lessons of a Feminine Science." *Isis* 84 (1993): 470–95.

Krell, David Farrell. *Contagion: Sexuality, Disease, and Death in German Idealism and Romanticism*. Bloomington: Indiana University Press, 1998.

Kuhn, Dorothea. "Goethe's Relationship to the Theories of Development of His Time." *Goethe and the Sciences: A Reappraisal*. Ed. Frederick Amrine, Francis J. Zucker, and Harvey Wheeler. Boston Studies in the Philosophy of Science, vol. 97. Boston: D. Reidel, 1987: 3–15.

Kuhn, Thomas S. *The Structures of Scientific Revolutions*. 3d ed. Chicago: University of Chicago Press, 1996.

Land, Edwin. "Color Vision and Natural Image." *Proceedings of the National Academy of Sciences* 45 (1959): 115–29.

————. "Experiments in Color Vision." *Scientific American* 200 (1959): 84–99.

Lauxtermann, P. F. H. "Hegel and Schopenhauer as Partisans of Goethe's Theory of Color." *Journal of the History of Ideas* 51 (1990): 599–624.

Lenoir, Timothy. "The Eternal Laws of Form: Morphotypes and the Conditions of Existence in Goethe's Biological Thought." *Goethe and the Sciences: A Reappraisal.* Ed. Frederick Amrine, Francis J. Zucker, and Harvey Wheeler. Boston Studies in the Philosophy of Science, vol. 97. Boston: D. Reidel, 1987: 17–28.

―――. *The Strategy of Life: Teleology and Mechanics in Nineteenth-Century Biology.* Chicago: University of Chicago Press, 1989.

Lewontin, Richard. *The Triple Helix: Gene, Organism, and Environment.* Cambridge: Harvard University Press, 2000.

Lillyman, William J. "Analogies for Love: Goethe's *Die Wahlverwandtschaften* and Plato's *Symposium*." *Goethe's Narrative Fiction.* Ed. William J. Lillyman. New York: Walter de Gruyter, 1982: 128–44.

Lönnig, Wolf-Ekkehard. "Goethe, Sex, and Flower Genes." *Plant Cell* 6 (1994): 574–76.

Magnus, Rudolf. *Goethe as Scientist.* Trans. Heinz Norden. New York: Henry Schuman, 1949.

Martin, Günther. "Goethes Wolkentheologie." *Zeitschrift für deutsche Philologie* 114 (1995): 182–98.

Mayr, Ernst. *The Growth of Biological Thought: Diversity, Evolution, and Inheritance.* Cambridge: Belknap of Harvard University Press, 1982.

Milfull, John. "The 'Idea' of Goethe's *Wahlverwandtschaften*." *Germanic Review* 47 (1972): 83–94.

Miller, J. Hillis. *Ariadne's Thread: Story Lines.* New Haven: Yale University Press, 1992.

―――. "A 'Buchstäbliches' Reading of *The Elective Affinities*." *Glyph* 6 (1979): 1–23.

―――. "Interlude as Anastomosis in *Die Wahlverwandtschaften*." *Goethe Yearbook* 6 (1992): 115–22.

Milton, John. *Paradise Lost. The Poetical Works of John Milton.* Vol. 1. Ed. Helen Darbishire. Oxford: Clarendon, 1952.

Müller, Gerhard. "*Wechselwirkung* in the Life and Other Sciences: A Word, New Claims and a Concept Around 1800 . . . and Much Later." *Romanticism in Science: Science in Europe 1790–1840.* Ed. Stefano Poggi and Maurizio Bossi. Dordrecht: Kluwer Academic Publishers, 1994: 1–14.

Müller-Sievers, Helmut. "Skullduggery: Goethe and Oken, Natural Philosophy and Freedom of the Press." *Modern Language Quarterly* 59 (1998): 231–59.

Muschg, Adolf. "'Im Wasser Flamme': Goethes grüne Wissenschaft." *Goethe als Emigrant: Auf der Suche nach dem Grünen bei einem alten Dichter.* Frankfurt am Main: Suhrkamp, 1986: 48–72.

Newton, Isaac. *Opticks or A Treatise of the Reflections, Refractions, Inflections & Colours of Light.* New York: Dover, 1979.

Nietzsche, Friedrich. *The Genealogy of Morals: An Attack.* Trans. Francis Golffing. New York: Doubleday Anchor, 1956.

―――. *Zur Genealogie der Moral: Eine Streitschrift.* Vol. 5 of *Kritische Studienausgabe.* 15

vols. Ed. Giorgio Colli and Mazzino Montinari. Munich: Deutscher Taschenbuch, 1988.

Nisbet, H. B. *Goethe and the Scientific Tradition.* London: Institute of Germanic Studies—University of London, 1972.

Noddings, Nel. *Caring: A Feminine Approach to Ethics and Moral Education.* Berkeley: University of California Press, 1984.

———. *Women and Evil.* Berkeley: University of California Press, 1989.

Nygaard, Loisa. "'Bild' and 'Sinnbild': The Problem of the Symbol in Goethe's *Wahlverwandtschaften.*" *Germanic Review* 63 (1988): 58–76.

Overbeck, Gertrud. "Goethes Lehre von der Metamorphose der Pflanzen und ihre Widerspiegelung in seiner Dichtung." *Publications of the English Goethe Society* 31 (1960–61): 38–59.

Pinto-Correia, Clara. *The Ovary of Eve: Egg and Sperm and Preformation.* Chicago: University of Chicago Press, 1997.

Plato. *The Collected Dialogues of Plato.* Ed. Edith Hamilton and Huntington Cairns. Princeton: Princeton University Press, 1980.

Pörksen, Uwe. "'Alles ist Blatt': Über Reichweite und Grenzen der naturwissenschaftlichen Sprache und Darstellungsmodelle Goethes." *Berichte zur Wissenschaftsgeschichte* 11 (1988): 133–48.

Portmann, Adolf. "Goethe and the Concept of Metamorphosis." *Goethe and the Sciences: A Reappraisal.* Ed. Frederick Amrine, Francis Zucker, and Harvey Wheeler. Boston Studies in the Philosophy of Science, vol. 97. Boston: D. Reidel, 1987: 133–45.

Prandi, Julie D. *"Dare to Be Happy": A Study of Goethe's Ethics.* Lanham, Md.: University Press of America, 1993.

Reiss, Hans. *Goethe's Novels.* New York: St. Martin's Press, 1969.

Ribe, Neil M. "Goethe's Critique of Newton: A Reconsideration." *Studies in History and Philosophy of Science* 16 (1985): 315–35.

Richter, Simon. *Laocoon's Body and the Aesthetics of Pain: Winckelmann, Lessing, Herder, Moritz, Goethe.* Detroit: Wayne State University Press, 1992.

Riegner, Mark, and John Wilkes. "Flowforms and the Language of Water." *Goethe's Way of Science: A Phenomenology of Nature.* Ed. David Seamon and Arthur Zajonc. Albany: SUNY Press, 1998: 233–52.

Roe, Shirley A. *Matter, Life, and Generation: Eighteenth-Century Embryology and the Haller-Wolff Debate.* Cambridge: Cambridge University Press, 1981.

Rousseau, Jean-Jaques. *Emile or On Education.* Trans. Allan Bloom. New York: Basic, 1979.

Rowland, Herbert. "Chaos and Art in Goethe's *Novelle.*" *Goethe Yearbook* 8 (1996): 93–107.

Sacks, Oliver. "The Case of the Colorblind Painter." *An Anthropologist on Mars: Seven Paradoxical Tales.* New York: Alfred A. Knopf, 1995: 3–41.

Schaeder, Grete. *Gott und Welt: Drei Kapitel Goethescher Weltanschauung.* Hameln: Fritz Seifert, 1947.

Schelling, Friedrich Wilhelm Joseph von. *Ideas for a Philosophy of Nature*. Trans. Errol E. Harris and Peter Heath. Cambridge: Cambridge University Press, 1988.

———. *Ideen zu einer Philosophie der Natur (1797)*. Vol. 5 of *Werke*. Ed. Manfred Durner. Stuttgart: Friedrich Frommann, 1994.

Schiebinger, Londa. *Nature's Body: Gender in the Making of Modern Science*. Boston: Beacon, 1993.

Schmidt, Alfred. *Goethes herrlich leuchtende Natur: Philosophische Studie zur deutschen Spätaufklärung*. Munich: Carl Hanser, 1984.

Schöne, Albrecht. *Goethes Farbentheologie*. Munich: C. H. Beck, 1987.

Schönherr, Hartmut. *Einheit und Werden: Goethes Newton-Polemik als systematische Konsequenz seiner Naturkonzeption*. Würzburg: Königshausen und Neumann, 1993.

Schrimpf, Hans Joachim. "Über die geschichtliche Bedeutung von Goethes Newton-Polemik und Romantik-Kritik." *Der Schriftsteller als öffentliche Person: Von Lessing bis Hochhuth*. Berlin: Erich Schmidt, 1977: 126–43.

Sepper, Dennis. "Goethe Against Newton: Towards Saving the Phenomenon." *Goethe and the Sciences: A Reappraisal*. Ed. Frederick Amrine, Francis J. Zucker, and Harvey Wheeler. Boston Studies in the Philosophy of Science, vol. 97. Boston: D. Reidel, 1987: 175–93.

———. *Goethe Contra Newton: Polemics and the Project for a New Science of Color*. Cambridge: Cambridge University Press, 1988.

Shteir, Ann. *Cultivating Women, Cultivating Science: Flora's Daughters and Botany in England 1760–1860*. Baltimore: Johns Hopkins University Press, 1996.

Sloan, Phillip R. "Organic Molecules Revisited." *Buffon* 88 (1992): 415–38.

Staiger, Emil. *Goethe*. Vol. 2. Zurich: Artemis, 1956.

Stephenson, R. H. *Goethe's Conception of Knowledge and Science*. Edinburgh: Edinburgh University Press, 1995.

Stern, Megan. "*Gravity's Rainbow* and the Newton/Goethe Colour Controversy." *Science as Culture* 4 (1993): 244–60.

Stern, Robert. Introduction to *Ideas for a Philosophy of Nature*. By Friedrich Wilhelm Joseph von Schelling. Trans. Errol E. Harris and Peter Heath. Cambridge: Cambridge University Press, 1988: ix–xxiii.

Tantillo, Astrida Orle. "Deficit Spending and Fiscal Restraint: Balancing the Budget in *Die Wahlverwandtschaften*." *Goethe Yearbook* 7 (1994): 40–61.

———. "Goethe's Botany and His Philosophy of Gender." *Eighteenth-Century Life* 22 (1998): 123–38.

———. "Goethe's Evolutionary Thinking." *Goethe, Chaos, and Complexity*. Ed. Herbert Rowland. Amsterdam: Rodopi, 2001: 47–56.

———. "Polarity and Productivity in Goethe's *Die Wahlverwandtschaften*." *Seminar* 36 (2000): 310–25.

Teller, Jürgen. "Totalität, Polarität, Steigerung, Menschenbezug: Grundbegriffe von Goethes Naturauffassung." *Goethe und die Wissenschaften*. Ed. Helmut Brandt. Jena: Friedrich-Schiller-Universität, 1984: 128–39.

Unseld, Siegfried. *Goethe and His Publishers*. Trans. Kenneth J. Northcott. Chicago: University of Chicago Press, 1996.

Vazsonyi, Nicholas. "Searching for 'The Order of Things': Does Goethe's *Faust II* Suffer from the 'Fatal Conceit'?" *Monatshefte* 88 (1996): 83–94.

Wachsmuth, Andreas B. "Bildung und Wirkung. Die Polarität in Goethes Lebenskunst." *Geeinte Zwienatur: Aufsätze zu Goethes naturwissenschaftlichem Denken*. Berlin: Aufbau, 1966: 86–112.

Waenerberg, Annika. *Urpflanze und Ornament: Pflanzenmorphologische Anregungen in der Kunsttheorie und Kunst von Goethe bis zum Jugendstil*. Helsinki: Finnish Society of Arts and Letters, 1992.

Wagenbreth, Otfried. "Goethes Stellung in der Geschichte der Geologie." *Goethe und die Wissenschaften*. Ed. Helmut Brandt. Jena: Friedrich-Schiller-Universität, 1984: 59–77.

Wells, George A. *Goethe and the Development of Science 1750–1900*. Alphen aan den Rijn, The Netherlands: Sijthoff and Noordhoff, 1978.

Wenzel, Manfred. "Goethe und Darwin: Der Streit um Goethes Stellung zum Darwinismus in der Rezeptionsgeschichte der morphologischen Schriften." *Goethe Jahrbuch* 100 (1983): 145–58.

———. "Goethe und Darwin: Goethes morphologische Schriften in ihrem naturwissenschaftshistorischen Kontext." Ph.D. diss., Ruhr Universität Bochum, 1982.

———. "Goethes Morphologie in ihrer Beziehung zum darwinistischen Evolutionsdenken." *Medizinhistorisches Journal* 18 (1983): 52–68.

Wilkinson, Elizabeth. "The Poet as Thinker: On the Varying Modes of Goethe's Thought." *Goethe: Poet and Thinker. Essays by Elizabeth M. Wilkinson and L. A. Willoughby*. London: Edward Arnold, 1962: 133–52.

Zirnstein, Gottfried. "Zu den Auffassungen über die Variabilität der Organismen bei Johann Wolfgang von Goethe und Zeitgenossen." *Goethe und die Wissenschaften*. Ed. Helmut Brandt. Jena: Friedrich-Schiller-Universität, 1984: 42–51.

INDEX